Textbook of
Environmental Science and Technology
(*Revised Edition*)

Textbook of
Environmental Science and Technology
(*Revised Edition*)

M. Anji Reddy
Professor of Environment
Centre for Environment
Jawaharlal Nehru Technological University Hyderabad,
Kukatpally, Hyderabad, A. P., India.
(anjijaya@sify.com)

BSP BS Publications
An Imprint of **BSP Books Pvt. Ltd.**
4-4-309/316, Giriraj Lane,
Sultan Bazar, Hyderabad - 500 095

Textbook of Environmental Science and Technology (Revised Edition)
***by* Anji Reddy**

Eleventh Reprint 2023

Published by:

BS Publications
An Imprint of **BSP Books Pvt. Ltd.**
4-4-309/316, Giriraj Lane, Sultan Bazar,
Hyderabad - 500 095
Phone : 040 – 23445688
e-mail: info@bspbooks.net
www.bspbooks.net

Printed at
Aditya Offset Process (I) Pvt. Ltd.
Hyderabad.

Price: Rs. 395.00
ISBN: 978-93-81075-44-9 (Paperback)

ANDHRA PRADESH STATE COUNCIL OF HIGHER EDUCATION
(A Statutory body of the Govt. of A.P.)
Post Box No : 34, Opp. Mahavir Hospital, Mahavir Marg, Masab Tank, Hyderabad - 500 028.
✆ : (O) 040 - 23311879 Fax : 040 - 23311470
E-mail : kalluri_c@yahoo.com URL : www.apsche.org

Prof. K.C. Reddy
CHAIRMAN

Foreword

Environmental issues have come to occupy the center stage of the discussions, both academic and policy making in the recent past not only because of the growing imbalances in the environmental resources but more importantly because of the concern and importance attached to equity and sustainability as components of quality of life in the development strategy. Environmental Management which calls for managing both resources and the technology that exploits them has become the cornerstone of development strategy as the inter-generational equity is now well documented in the development process. Environmental concerns impact both developed and developing societies. However, the nature of problems differ. During the middle of the 20th Century, South Asian countries like India, Pakistan and Bangladesh had two major problems. The first was providing food for the huge population and the second was preventing water contaminated diseases like dihorrea, cholera, typhoid, dysentery, etc. The yearly rainfall, though among the highest in the world in Bangladesh and in West Bengal, India, was not potent enough to meet the growing needs of both agriculture and industry. Moreover, India and Bangladesh, with plenty of available surface water, did not have the necessary infrastructure for the preservation, distribution, and purification facilities. The overall watershed management was poor. The farmer had to plea desperately for the rains in order to grow a harvest. The annual rainfall allowing a single harvest in a year was not enough for the population and the situation would be even worse, if there was a drought. Such circumstances called for alternative remedies.

The need for sustainable development is key factor to the future of mankind. Continuing problems of pollution, loss of forest, solid waster disposal, degradation of environment, issues like economic productivity and national security, Global warming, the depletion of ozone layer and loss of biodiversity have made everyone aware of environmental issues.

Inspite of the deteriorating status of the environment, detailed and critical study of environment has not received adequate attention in the academic programmes so far. Creating awareness among the youth on environmental concerns is now regarded extremely important in the long term interest of society. Recognizing this, the Hon'ble Supreme Court directed the UGC to introduce a basic course on environment at college level education. The matter was, therefore, considered by the UGC and it was decided that a six month compulsory core module course in environmental studies be prepared and implemented compulsorily in all the Indians Universities and colleges.

The hope for the future of environment in India is based on many positive factors: rich biodiversity and other natural resources, increasing number of peoples movements focusing on environment, greater public and media concern for these issues and the spread of environmental awareness among children and youth and compulsory teaching of appropriate environmental courses at all levels of education.

The present effort by the author Dr.Anji Reddy in the form of a Textbook of Environmental Science and Technology is an important step towards achieving the goal of environmental education. I am sure that this publication, based on the UGC Syllabus, will prove to be of immense value and will serve the needs of evironmentalists, agriculturists, students and researchers. The author had put in sincere efforts in bringing together information on a wide array of environmental issues and I would like to compliment him for the same. I with him every success in future endeavors.

(K.C. REDDY)

(Chairman)

A.P. State Council of Higher Education

JAWAHARLAL NEHRU TECHNOLOGICAL UNIVERSITY HYDERABAD
(Established by Andhra Pradesh Act No.30 of 2008)

Dr. M. Anji Reddy
Professor of Env. Science & Tech.

Kukatpally
Hyderabad-500085
Andhra Pradesh (India)
Phone:040-23152329(O)
Fax: 040-23152330
e-mail:anjijaya@sify.com
www.jntu.ac.in

Preface

Importance of environmental protection, improvement of quality of life, conservation of biodiversity, solid waste management, and conservation of natural resources has been growing in the developed world in the later half of the twentieth century and this in turn has resulted in emphasis on environmental education around the world. United Nations has been a major force in this process. In India, the UGC guidelines on the syllabus of *Environmental Studies* may be considered as the logical culmination of the ongoing environmentalism around the world.

Inspite of the deteriorating status of the environment, study of environment has so far not received adequate attention in our academic programs. Recognizing this, the Hon'ble Supreme Court directed the UGC to introduce a basic course on Environmental Studies at every level of higher education. Accordingly, the matter was considered by all universities and it was decided that compulsory course in environmental studies be prepared and implemented in all the JNTUH affiliated engineering colleges. Keeping this in view, I have revised the book to meet the academic requirements of JNTUH and its affiliated engineering colleges. This book consists of ten chapters and each chapter ends with review questions that provide added practice for the students.

I am deeply grateful to faculty and reviewers who have suggested changes for this book. I am also indebted to those authors whose books are source of information. Special thanks to Mr. G.V. Reddy of Team labs, Mr. P. Raju, and Dr. Vijaya Kumar of Ramkey for providing necessary help in writing this book. I thank Mr. Nikhil Shah, Proprietor of BS Publications for timely publication of this book. Finally, special thanks to my wife Jaya and children Chandu and Rajiv for their support and cooperation in writing this book.

M. Anji Reddy

Contents

1

ECOSYSTEMS

2

NATURAL RESOURCES

8

Environmental Impact Assessment

9

TECHNOLOGICAL SOLUTIONS FOR POLLUTION CONTROL

10

TOWARDS SUSTAINABLE FUTURE

JAWAHARLAL NEHRU TECHNOLOGICAL UNIVERSITY HYDERABAD

II Year B.Tech

ENVIRONMENTAL STUDIES

UNIT-I : ECOSYSTEMS: Definition, Scope and Importance of ecosystem, Concept of ecosystem, Classification of ecosystems, Structure and Structural Components of an ecosystem, Functions of ecosystem, Food chains, food webs and ecological pyramids. Flow of energy, Biogeochemical cycles, Homeostasis / Cybernetics, Food chain concentration, Biomagnification, ecosystems value, services and carrying capacity.

UNIT-II: NATURAL RESOURCES: Classification of Resources: Living and Non-Living resources, Renewable and non-renewable resources. Water resources: use and over utilization of surface and ground water, floods and droughts, Dams: benefits and problems. Mineral resources: use and exploitation, environmental effects of extracting and using mineral resources – case studies. Energy resources: growing energy needs, renewable and non renewable energy sources, use of alternate energy sources – case studies. Land resources: land as a resource, land degradation, man induced landslides and land use / land cover mapping.

UNIT-III: BIODIVERSITY AND BIOTIC RESOURCES: Introduction, Definition, genetic, species and ecosystem diversity. Value of biodiversity: consumptive use, productive use, social, ethical, aesthetic and intrinsic values. Hot spots of biodiversity. Threats to biodiversity: habitat loss, poaching of wildlife, man-wildlife conflicts, conservation of biodiversity: In-Situ and Ex-situ conservation. Food and fodder resources, Timber and non-timber forest products.

UNIT-IV: ENVIRONMENTAL POLLUTION AND CONTROL: Classification of pollution and pollutants, causes, effects and control technologies. Air Pollution: Primary and secondary pollutants, Automobile and Industrial pollution, Ambient air quality standards. Water pollution: Point and non-point sources of pollution, Major pollutant of water and their sources, drinking water quality standards, Waste water treatment methods: effluent treatment plants (ETP), Sewage treatment plants (STP), common and combined effluent treatment plants (CETP).Soil Pollution: Soil as sink for pollutants, Impact of modern agriculture on soil, degradation of soil. Marine Pollution: Misuse of International water for dumping of hazardous waste, coastal pollution due to sewage and marine disposal of industrial effluents. Noise Pollution: Sources, Industrial Noise- Occupational Health hazards, standards, Methods of control of Noise. Thermal Pollution: Thermal Comforts, Heat Island effect, Radiation effects. Nuclear Pollution: Nuclear power plants, nuclear radiation, disasters and impacts, genetical disorders. Solid waste: types, Collection processing and disposal of industrial and municipal solid wastes composition and characteristics of e-Waste and its management.

UNIT-V: GLOBAL ENVIRONMENTAL PROBLEMS AND GLOBAL EFFORTS : Green house effect, Green House Gases (GHG), Global Warming, Sea level rise, climate change and their impacts on human environment. Ozone depletion and Ozone depleting substances (ODS).Deforestation and desertification. International conventions / Protocols: Earth summit, Kyoto protocol and Montréal Protocol,

UNIT-VI: ENVIRONMENTAL IMPACT ASSESSMENT (EIA) AND ENVIRONMENTAL MANAGEMENT PLAN: Definition of Impact: classification of impacts, Positive and Negative, Reversible and irreversible, light, moderate and severe, methods of baseline data acquisition. Impacts on different components: such as human health resources, air, water, flora, fauna and society. Prediction of impacts and impact assessment methodologies. Environmental Impact Statement (EIS). Environmental Management Plan (EMP): Technological Solutions, preventive methods, Control technologies, treatment technologies: green-belt-development, rain water harvesting, Remote sensing and GIS methods.

UNIT-VII: ENVIRONMENTAL POLICY, LEGISLATION, RULES AND REGULATIONS

National Environmental Policy, Environmental Protection act, Legal aspects Air (Prevention and Control of pollution) Act- 1981, Water(Prevention and Control of pollution) Act-1974, Water pollution Cess Act-1977, Forest Conservation Act, Municipal solid waste management and handling rules, biomedical waste management and handling rules, hazardous waste management and handling rules .

UNIT: VIII -- TOWARDS SUSTAINABLE FUTURE

Concept of Sustainable Development, Threats to Sustainability, Population and its explosion, Crazy Consumerism, Over-exploitation of resources, Strategies for Achieving Sustainable development, Environmental Education, Conservation of Resources, Urban Sprawl, Sustainable Cities and Sustainable Communities, Human health, Role of IT in Environment, Environmental Ethics, Environmental Economics, Concept of Green Building, Clean Development Mechanism (CDM).

CHAPTER 1

ECOSYSTEMS

1.1 DEFINITION

An ecosystem is a segment of nature consisting of a community of living beings and the physical environment both interacting and exchanging materials between them. The two components of nature, organisms and the environment are not only highly complex and dynamic but also interdependent, mutually reactive and interrelated. Ecology deals with the various principles which govern relationships between organisms and their environment. According to the French Zoologist, Isodore Geoffroy St. Hilaire, the term ecology is used for the study of the relations of the organisms within the family and society and community. At about the same time the English naturalist, St. George Jackson Mivart coined the term hexicology, and defined as the study of the relations which exist between the organisms and their environment with respect to the nature of the locality, the temperature and the amounts of light, and their relations to other organisms as enemies, rivals, or accidental and involuntary benefactors. Thus, ecology is literally the study of organisms and their relationship with environment.

1.2 SCOPE AND IMPORTANCE

Taylor (1936), in an attempt to define ecology, has very rightly pointed out that scope of ecology by stating that ecology is the science of all the relations of ecosystems, all organisms to all their environments. Ecology plays an important role in agriculture crop rotation, weed control, management of grasslands, range management forestry, biological surveys, pest control, fishery biology, and in the conservation of soil. wildlife, forest, water supplies, water bodies like rivers, lakes and ponds,

Ecosystem is defined as a dynamic entity composed of a biological community and its associated abiotic environment. Often the dynamic interactions that occur within an ecosystem are numerous and complex. Ecosystems are also always undergoing alterations to their biotic and abiotic components. Some of these alterations begin first with a change in the state of one component of the ecosystem, which then cascades and sometimes amplifies into other components because of relationships.

1.3 CONCEPTS OF ECOLOGY

Based on structural components and their relationships, ecology and ecosystem can be explained in two ways: structural concepts and functional concepts. In recent years, the impact of human has caused a number of dramatic changes to a variety of ecosystems found on the Earth. Humans use and modify natural ecosystems through agriculture, forestry, recreation, urbanization, and industry. The most obvious impact of humans on ecosystems is the loss of biodiversity. The number of extinctions caused by human domination of ecosystems has been steadily increasing since the start of the Industrial Revolution. The frequency of species extinctions is correlated to the size of human population on the Earth, which is directly related to resource cons☺umption, land-use change, and environmental degradation. Other human impacts to ecosystems include species invasions to new habitats, changes to the abundance and dominance of species in Communities, modification of biogeochemical cycles, modification of hydrologic cycling, pollution, and climatic change.

Interdependence is the important basis of ecology. Ecosystem includes interacting and interdependent components that are open and linked to each other. An ecosystem may be as small as a drop of pond water (micro ecosystem) or as large as ocean. It can be of temporary nature, for example a fresh pool, a field of cultivated crops, a forest or an ocean. A balanced aquarium may be thought of as an artificially established self-sustained ecosystem. It must be noted that a completely self-sufficient ecosystem like aquarium is rarely found in nature, rather, all these unit ecosystems are simply separated from each other with time and space, but functionally they all are, linked forming as integrated whole.

1.3.1 Structural Concepts

The different types of organisms living in a particular environment are not only independent and mutually reactive but also react with the environment. Though organisms of a species maintain uniformity in their structure and functions through having a common gene pool, they have sufficient plasticity to modify themselves according to changing environment by modifications in somatic characters (ecads) or genetic characters (ecotypes). Due to their activities, organisms modify the environment to make it more congenial for their growth, development, reproduction and dispersal. The modified environment may become less suitable for the community already living in it. This invites another community that also changes the environment may become less suitable for the community already living in it. This invites another community that also changes the environment further beyond its most favorable limit. The development of different communities over a period of time at the same site is called succession. The process of succession and change in environment would continue till an equilibrium is established between the changed environment and a community called climax community. Under similar climatic conditions, different types of communities grow. Some of them have reached their climax stage while others occur in different stages of succession. The complex of many communities growing in a particular area and sharing a common climate is called biome.

1.3.2 Functional Concepts

The biological community consists of a number of organisms and/or populations. Each population occupies a specific volume of the habitat circumscribed by the interaction of various environmental factors and trophic level of the organisms. It is called ecological niche. The degree of success of a particular population in an area is determined by the parameters of both abiotic factors as well as interaction with other types of populations. The interactions amongst the populations can be positive, negative or neutral.

The flow of energy in the ecosystem is unidirectional or non-cyclic. Radiant energy is trapped by autotrophic plants or primarily producers. From there the energy is transferred to consumers and decomposers Energy is lost during its transfer from one trophic level to the next. Organisms use the energy in respiration.

A number of inorganic substances are taken by the living beings for their metabolism and body building. They are called biogenetic nutrients. The biogenetic nutrients keep on circulating between the biotic and abiotic components of the ecosystem. The phenomenon is called biogeochemical cycling. Human beings exploit the ecosphere for their own benefits. As a result, only the economically important plants are allowed to grow in an ecosystem. Species diversity and natural interactions amongst the various components are reduced. When neglected, such an ecosystem deteriorates.

A disturbed or deteriorated ecosystem shows changes due to interactions inside, the assemblage of living being and their abiotic environment, modifying and changing both abiotic and biotic components. The change continues till a stable climax community develops. Where a disturbance continues, the deteriorated ecosystem changes the environment completely.

1.4 CONCEPT OF AN ECOSYSTEM

It shows that earth itself is a vast ecosystem (biosphere) where abiotic and biotic components are constantly acting and reacting upon each other. The biosphere has various smaller ecosystems, for example forest, desert, grassland, cropland, freshwater, marine, etc. Ecosystem is usually an open system with a continuous, but variable influx and loss of materials and energy. An ecosystem represents the highest level of ecological integration, which is energy based and capable of energy transformation, accumulation and circulation. Its main function in ecological sense is to emphasize obligatory relationship, interdependence and casual relationships, that is coupling of components to form functional units. Ecosystem may also be regarded as energy processing unit which is restrained or limited by the amount of nutrients and water available to it. An ecosystem is thus an overall integration of white mosaic of interacting organisms and their environment. It is normally an open system with a continuous, but variable influx and loss of materials and energy

1.5 MAJOR COMPONENTS OF AN ECOSYSTEM

Ecosystems are composed of a variety of abiotic and biotic components that function in an interrelated fashion. Some of the more important components are: soil, atmosphere, radiation from the sun, water, and living organisms. Soils are much more complex than simple sediments. They contain a mixture of weathered rock fragments, highly altered soil mineral particles, organic matter, and living organisms. Soils provide nutrients, water, a home, and a structural growing medium for organisms. The vegetation found growing on top of a soil is closely linked to this component of an ecosystem through nutrient cycling. The atmosphere provides organisms found within ecosystems with carbon dioxide for photosynthesis and oxygen for respiration. The processes of evaporation, transpiration. and precipitation cycle water between the atmosphere and the Earth's surface.

Solar radiation is used in ecosystems to heat the atmosphere and to evaporate and transpire water into the atmosphere. Sunlight is also necessary for photosynthesis. Photosynthesis provides the energy for plant growth and metabolism, and the organic food for other forms of life. Ecosystems are composed of a variety of living organisms that can be classified as producers, consumers, or decomposers. Producers or autotrophs, are organisms that can manufacture the organic compounds they use as sources of energy and nutrients. Most producers are green plants that can manufacture their food through the process of photosynthesis. Consumers or heterotrophs get their energy and nutrients by feeding directly or indirectly on producers. We can distinguish two main types of consumers. Herbivores are consumers that eat plants for their energy and nutrients. Organisms that feed on herbivores are called carnivores. Carnivores can also consume other carnivores. Plants and animals supply organic matter to the soil system through shed tissues and death. Consumer organisms that feed on this organic matter, or detritus, are known as detritivores or decomposers. The organic matter that is consumed by the detritivores is eventually converted back into inorganic nutrients in the soil. These nutrients can then be used by plants for the production of organic compounds.

1.6 STRUCTURE AND FUNCTION OF AN ECOSYSTEM

The two major aspects of an ecosystem are the structure and function. By structure we mean (i) the composition of biological community including species, numbers, biomass, life history and distribution in space etc., (ii) the quantity and distribution of the non-living materials, such as nutrients, water etc., and (iii) the range, or gradient of conditions of existence, such as temperature, light etc. By function we mean (i) the rate of biological energy flow i.e., the production and respiration rates of the community, (ii) rate of materials or nutrient cycles, and (iii) biological or ecological regulation including both regulation of organisms by environment (photoperiodism etc.) and regulation of environment by the organism, (nitrogen fixing organisms etc.). Thus, in any ecosystem, structure and function (rate functions) are studied together.

1.6.1 Structure of an Ecosystem

An ecosystem has two major components: namely abiotic and biotic. Abiotic (non-living) component includes

(i) The amount of inorganic substances as P, S, C, N, H, etc. involved in material cycles. The amount of these inorganic substances, present at any given time in an ecosystem, is designated as the standing state or standing quality,

(ii) Amount and distribution of inorganic chemicals, such as chlorophylls etc., and of organic materials, as proteins, carbohydrates, lipids etc., present either in the biomass or in the environment i.e. biochemical structure that link the biotic and abiotic components of the ecosystem,

(iii) the climate of the given region.

The Biotic (living) component is indeed the trophic structure of any ecosystem, where living organisms are distinguished on the basis of their nutritional relationships. Biotech components of an ecosystem has two sub components: namely autotrophic and heterotrophic.

(i) **Autotrophic component.** In which fixation of light energy, use of simple inorganic substances and build up of complex substances predominate. The component is constituted mainly by green plants, including photosynthetic bacteria. To some lesser extent, chemosynthetic microbes also contribute to the build up of organic matter. Members of the autotrophic component are known as producers.

(ii) **Heterotrophic component.** In which utilisation, rearrangement and decomposition of complex materials predominate. The organisms involved are known as consumers, as they consume the matter built up by the producers (autotrophs). The consumers are further categorised as: macro and micro consumers.

(a) **Macro consumers:** These are the consumers, which in an order as they occur in a food chain are, herbivores, carnivores (or omnivores). Herbivores are also known as primary consumers. Secondary and tertiary consumers, if present, are carnivores or omnivores. They all are phagotrophs which include chiefly animals that ingest other organic and particulate organic matter.

(b) **Micro consumers:** These are popularly known as decomposers. They are saprotrophs (osmotrophs) and include chiefly bacteria, actinomycetes and fungi. They breakdown complex compounds of dead or living protoplasm absorb some of the decomposition or breakdown products and release inorganic nutrients in environment, making them available again to autotrophs.

The biotic component of any ecosystem may be thought of as the functional kingdom of nature, since they are based on the type of nutrition and the energy source used. The trophic structure of an ecosystem is one kind of producer consumer arrangement, where each "food" level is known as trophic level. The amount of living material in different trophic levels or in a component population is known as the standing crop, a term applicable to both, plants as well as animals. The standing crop may be expressed in terms of (i) number of organisms per unit area, or (ii) biomass i.e. organism mass in unit area, which can be measured as living weight, dry weight, ash-free dry weight or carbon weight, or calories or any other convenient unit suitable for comparative purposes.

1.6.2 Function of an Ecosystem

Many of the most important relationships between living organisms and the environment are controlled ultimately by the amount of available incoming energy received at the Earth's surface from the sun. It is this energy, which helps to drive biotic systems. The sun's energy allows plants to convert inorganic chemicals into organic compounds. Only a very small proportion of the sunlight received at the Earth's surface is transformed into biochemical form. Living organisms can use energy in basically two forms: radiant or fixed. Radiant energy exists in the form of electromagnetic energy, such as light. Fixed energy is the potential Chemical energy found in organic substances. This energy can be released through respiration. Organisms that can take energy from inorganic sources and fix it into energy rich organic molecules are called autotrophs. If this energy comes from light then these organisms are called photosynthetic autotrophs. In most ecosystems plants are the dominant photosynthetic autotroph. Organisms that require fixed energy found in organic molecules for their survival are called heterotrophs. Heterotrophs who obtain their energy from living organisms are called consumers. Consumers can be of two basic types: Consumer and decomposers. Consumers that consume plants are known as herbivores. Carnivores are consumers who eat herbivores or other carnivores. Decomposers or detritivores are heterotrophs that obtain their energy either from dead organisms or from organic compounds dispersed in the environment.

The behaviour of energy in ecosystem can be termed energy flow due to unidirectional flow of energy, From energetics point of view it is essential to understand for an ecosystem (i) the efficiency of the producers in absorption and conversion of solar energy, (ii) the use of this converted chemical form of energy by the consumers, (iii) the total input of energy in the form of food and its efficiency of assimilation, (iv) the loss through respiration, heat, excretion etc. and (v) the gross net production. Two energy models to understand typical ecosystem. They are single channel energy models and g-shaped energy flow models.

1.7 FOOD CHAIN, FOOD WEBS & ECOLOGICAL PYRAMIDS

1.7.1 Food Chains

In nature, we generally distinguish two general types of food chains: of razing food chain and detritus food chain. The Grazing food chain starts from the **living** green plants, goes to grazing herbivores (that feed on living plant materials with their predators), and on to carnivores (animal eaters). Ecosystems with such type of food chain are directly dependent on an influx of solar radiation. This type of chain thus depends on autotrophic energy capture and the movement of this captured energy to herbivores. Most of the ecosystems in nature follow this type of food chain. From energy standpoint, these chains are very important. The phytoplanktons - zooplanktons - fish sequence or the grasses – rabbit- fox sequence are the examples of grazing food chain.

The Detritus Food chain goes from dead organic matter into microorganisms and then to organisms feeding on detritus (detritivores) and their predators. Such ecosystems are thus less dependent on direct solar energy. These depend chiefly on the influx of organic matter produced in another system. A good example of a detritus food chain is based on mangrove leaves. All the animals are detritus consumers. These detritivores are the key group of small animals, comprising only a few species but very large number of individuals. They ingest large amounts of the vascular plant detritus. These animals are in turn eaten by some minnows and small game fish etc. that is the small carnivores, which in turn serve as the main food for larger game fish and fish eating birds which are the large (top) carnivores. The mangroves considered generally as of less economic value make a substantial contribution to the food chain that supports the fisheries, an important economy in that region. Similarly detritus from sea grasses, salt marsh grasses and seaweeds support fisheries in many estuarine areas.

Thus the detritus food chain ends up in a manner similar to the grazing food chain (big fish eat little fish), but the way in which the two chains begin is quite different. In detritus chain, the detritus consumers, in contrast to grazing herbivores, are a mixed group in terms of trophic levels. These include herbivores, omnivores and primary carnivores. As a group, the detritus feeders obtain some of their energy directly from plant material, most of it secondarily from microorganisms, and some tertiarily through carnivores (for example by eating protozoa or other small invertebrates that have fed or bacteria that have digested plant material).

But under natural situations, system must always be self sufficient. In fact this type of food chain (detritus type) is simply a sub-component of another ecosystem. And, the above said two types of food chain in nature are indeed linked together belonging to the same ecosystem.

1.7.2 Food Webs

Food chains in natural conditions never operate as isolated sequences, but are interconnected with each other forming interlocking pattern, which is referred to as a food web. Under natural conditions, the linear arrangement of food chains hardly occurs and these remain indeed interconnected with each other through different types of organisms at different trophic levels. For example, in grazing food chain of grassland, in the absence of rabbit, grass may also be eaten by mouse. The mouse in turn may be eaten directly by hawk or by snake first which is then eaten by hawk. Thus, in nature there are found alternatives, which all together constitute some sort of interlocking pattern and is called the food web. There may be seen five possible food chains interlocked together making the food web. Besides those shown in Figure 1.1, there may also be present some other consumers as vultures, fox and man in grasslands, and if so, the food web may be even more complex than shown in the figure. However, these all five chains are interlinked with each other at different points, forming food web. Real food webs usually have hundreds of species interlinked by their feeding habits. Food webs are basic units of ecosystem ecology.

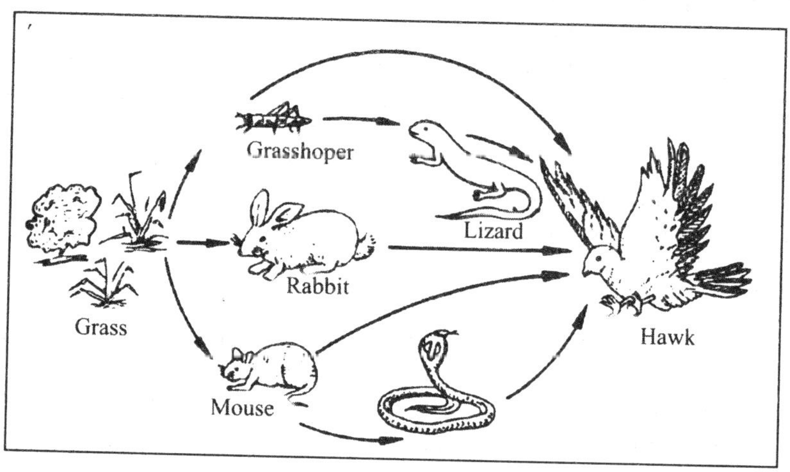

Fig. 1.1 Diagrammatic sketch showing a food web in a grassland ecosystem.

A balanced ecosystem is essential for the survival of all living organisms of the system. For instance, had primary consumers (herbivores) not been in nature producers would have perished due to overcrowding and competition. Similarly, the survival of primary consumers is linked with the consumers (carnivores) and so on. Thus, each species of any ecosystem is indeed kept under some sort of a natural check so that the system may remain balanced. The complexity of any food web depends upon the diversity of organisms in the system. It would accordingly depend upon two main points: (i) length of the food chain. Diversity in the organisms based upon their food habits would determine the length of food chain. More diverse the organisms in food habits, more longer would be food chain, (ii) alternatives at different points of consumers in the chain. More the alternatives more would be the interlocking pattern. In deep oceans, seas. where we find a variety of organisms, the food webs are much complex.

1.7.3 Ecological Pyramids

Each ecosystem is characterized by the interaction of food chain, the metabolism relationship between the linearly arranged biotic components of an ecosystem. This characteristic is called the trophic structure. The trophic structure and function at successive trophic levels, that is producers → herbivores → carnivores, may be shown graphically by means of ecological pyramids where the first or producer level constitutes the base of the pyramid and the successive levels, the tiers making the apex. Ecological pyramids are of three general types namely (i) pyramid of numbers, showing the number of individual organisms at each level (ii) pyramid of biomass, showing the total dry weight and other suitable measure of the total amount of living matter, and (iii) pyramid of energy, showing the rate of energy flow and/or productivity at successive trophic levels. The pyramids of numbers and biomass may be upright or inverted depending upon the nature of the food chain in the particular ecosystem, whereas pyramids of energy are always upright.

1.7.3.1 *Pyramids of Numbers*

Pyramids of numbers show the relationship between producers, herbivores and carnivores at successive trophic levels in terms of their number. In a grassland the producers, which are mainly grasses are always maximum in number. This number then shows a decrease towards apex, as the primary consumers (herbivores) like rabbits, mice etc. are lesser in number than the grasses; the secondary consumers, snakes and lizards are lesser in number than the rabbits and mice. Finally, the top (tertiary) consumers hawks or other birds are least in number. Thus, the pyramid becomes upright. Similarly, in a pond ecosystem the pyramid is upright. Here the producers which are mainly the phyto-planktons as algae, bacteria etc. are maximum in number The herbivores, which are smaller fish rotifers etc. are lesser in number than producers and the secondary consumers (carnivores), such as small fish eating each other, water beetles etc. are lesser in number than the herbivores. Finally, the top (tertiary) consumers, the bigger fish are least in number. (In a forest ecosystem, however, the pyramid of numbers is somewhat different in shape. The producers, which are mainly large-sized trees, are lesser in number, and form the base of the pyramid. The herbivores, which are the fruit-eating birds, elephants, deers etc.are more in number than the producers. Then there is a gradual decrease number of successive carnivores, thus making the pyramid again upright. However in a parasitic food chain the pyramids are always inverted. This is due to the fact that a single plant may support the growth of many herbivores and each herbivore in turn may provide nutrition to several parasites which support many hyperparasites. Thus, from the producer towards consumers, there is a reverse position, i.e. the number of organisms gradually shows an increase making the pyramid inverted in shape. Actually the pyramids of numbers do not give a true picture of the food chain as they are not very functional. They do not indicate the relative effects of the 'geometric', 'foodchain' and 'size' factors of the organisms. (Fig. 1.2)

1.7.3.2 *Pyramids of Biomass*

They are comparatively more fundamental, instead of geometric factor show the quantitative relationships of the standing crops. In grassland and forest, there is generally a gradual decrease in biomass of organisms at successive levels from the producers to the top carnivores. Thus pyramids are upright. However, in a pond as the producers are small organisms, their biomass is least, and this value gradually shows an increase towards the apex of the pyramid, thus making the pyramid inverted in shape.

1.7.3.3 *Pyramid of Energy*

Of the three types of ecological pyramids the energy pyramids give the best picture of overall nature of the ecosystem. Here number and weight of organisms at any level depends not on the amount of fixed energy present at any one time in the level just below but rather on the rate at which food is being produced. In contrast with the pyramids of numbers and biomass, which are pictures of the standing situations like organisms present at any moment the pyramid of energy is a picture of the rates of passage of food, mass through the food chain. In shape it is always upright, as in most of the cases there is always a gradual decrease in the energy content at successive trophic levels from the producers to various consumers.

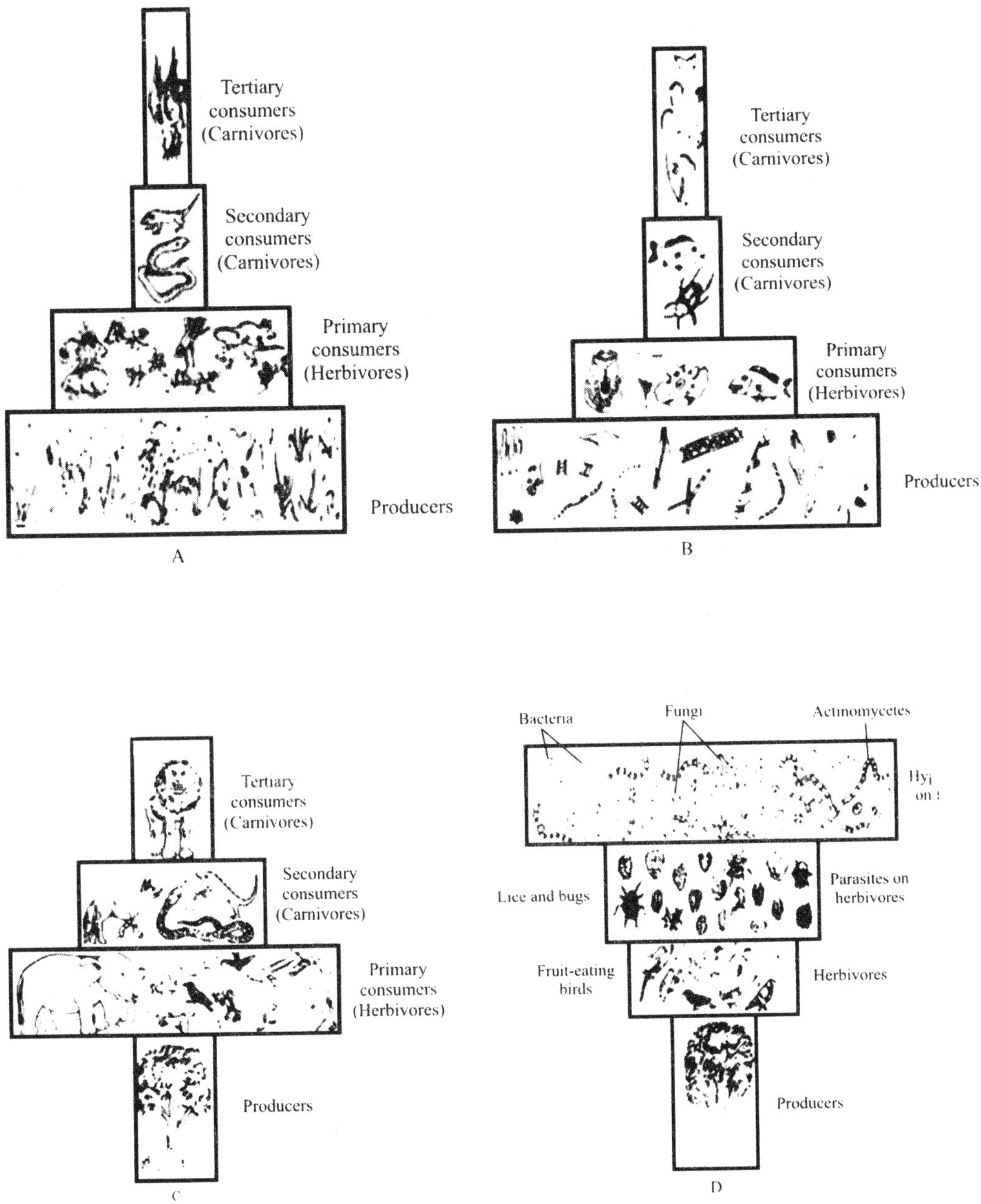

Fig. 1.2 Ecological pyramids

1.8 FLOW OF ENERGY

Flow of Energy can be explained by means of two models namely: single channel energy model and Y-shaped energy model.

1.8.1 Single-Channel Energy Models

The principle of food chains and the working of the two laws of thermodynamics can be better made clear by means of, energy flow diagrams shown in Figures 3 and 4. As shown in Figure 3 out of the total incoming solar radiation (118,872 gcal/cm2/yr), 118,761 gcal/cm2/yr remain unuttilised, and thus gross production (net production plus respiration) by autotrophs is 111 gcal/cm2/yr with an efficiency of energy capture of 0.1 0 per cent. It may also be noted that 21 percent of this energy or 23 gcal/cm2/yr is consumed in metabolic reactions of autotrophs for their growth, development, maintenance and reproduction.

It may be seen further that 15 gcal/cm2/yr are consumed by herbivores that graze or feed on Autotrophs--this amounts to 17 per cent of net autotroph production

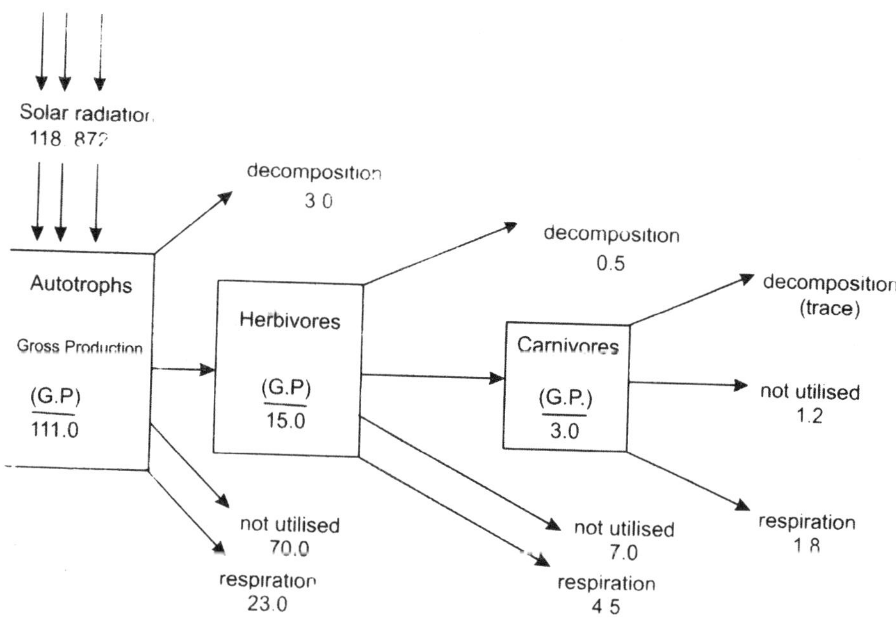

Fig. 1.3 Energy flow diagram for a lake (freshwater ecosystem) in g cal/cm2/yr

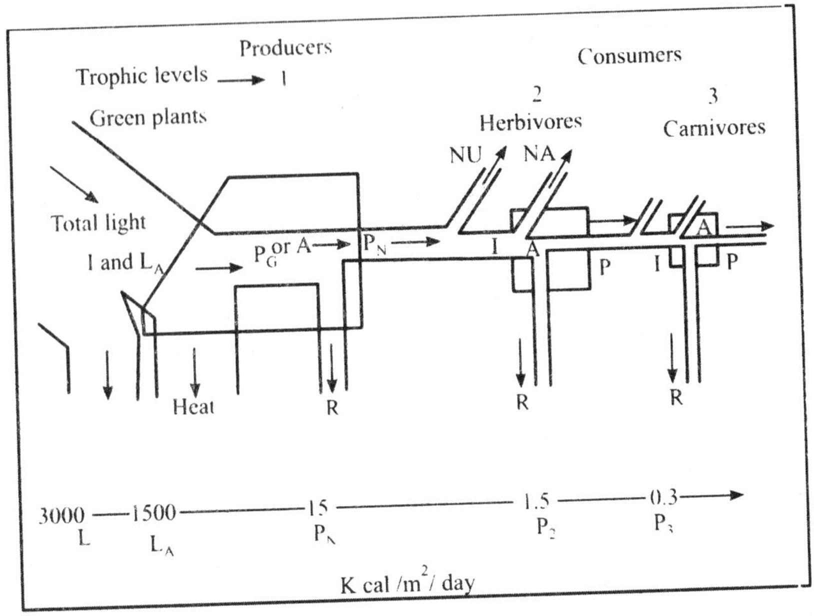

Fig. 1.4 A simplified energy flow diagram depicting three trophic levels

Decomposition (3 gcal/cm^{2}yr) accounts for about 3.4 per cent of net production. The remainder of the plant material, 70 gcal/cm^2/yr or 79.5 per cent of net production, is not utilised at all but becomes part of the accumulating sediments. It is obvious, then that much more energy is available for herbivory than is consumed. It may also be noted that various pathways of loss are equivalent to and account for energy capture of the autotrophs i.e. gross production. Also, collectively the three upper 'fates' (decomposition, herbivory and not utilised) are equivalent to net production. of the total energy incorporated at the herbivores level, i.e. 15 gcal.cm^2/yr, 30 percent or 4.5 gcal/cm^2/yr is used in metabolic reactions. Thus, there is considerably more energy lost via respiration by herbivores (30 percent) than by autotrophs (21 per cent).

Again there is considerable energy available for the carnivores, namely 10.5 gcal/cm^2/yr or 70 per cent, which is not entirely utilised; in fact only 3.0 gcal/cm^2/yr or 28.6 per cent of net production passes to the carnivores. This is more efficient utilisation of resources than occurs at autotroph-herbivore transfer level. At the carnivore level about 60 percent of the carnivores' energy intake is consumed in metabolic activity and the remainder becomes part of the not utilised sediments; only an insignificant amount is subject to decomposition yearly. This high respiratory loss compares with 30 per cent by herbivores and 21 per cent by autotrophs in this ecosystem. From the energy flow diagram shown in Figure 3, two things become clear. Firstly, there is one-way street along which energy moves (unidirectional flow *of* energy). The energy that is captured by the autotrophs does not revert back to solar input; that which passes to the herbivores does not pass back to the autotrophs. As it moves progressively through the various trophic levels it is no longer available to the Previous level. Thus due to one-way flow *of* energy, the system would collapse if the primary source, the sun, were cut off.

Secondly, there occurs a progressive decrease in energy level at each trophic level. This is accounted largely by the energy dissipated as heat in metabolic activities and measured here as respiration coupled, with unutilised energy. In Figure above the "boxes" represent the trophic levels and the 'pipes' depict the energy flow in and out *of* each level. Energy inflows balance outflows as required by the first law *of* thermodynamics, and energy transfer is accompanied by dispersion *of* energy into unavailable heat (i.e. respiration) as required by the second law. Figure 4 presents a very simplified energy flow model *of* three tropic levels, from which it becomes evident that the energy flow is greatly reduced at each successive trophic level from producers to herbivores and then to carnivores. Thus at each transfer *of* energy from one level to another, major part of energy is lost as heat or other form. There is a successive reduction in energy flow whether we consider it in terms *of* total flow (i.e. total energy input and total assimilation) *or* secondary production and respiration components. Thus, *of* the 3,000 Kcal of total light falling upon the green plants, approximately 50 per cent (1500Kcal) is absorbed, *of* which only 1 per cent (15 Kcal) is converted at first trophic level. Thus net primary production is merely 15 Kcal. Secondary productivity (P2 and P3 in the diagram) tends to be about 10 per cent at successive consumer trophic levels i.e. herbivores and the carnivores, although efficiency may be sometimes higher, as 20 per cent, at the carnivore level as shown (or P3= 0.3 Kcal) in the diagram.

It becomes evident from Figures 3 and 4 that there is a successive reduction in energy flow at successive trophic levels. Thus shorter the food chain, greater would be the available food energy as with an increase in the length *of* food chain there is a corresponding more loss *of* energy.

1.8.2 Y-shaped Energy Flow Models

The Y-shaped model further indicates that the two food chains namely the grazing food chain and detritus food chain are in fact, under natural conditions, not completely isolated from one another. The grazing food chain beginning with green plant base going to herbirores and the detritus food chain beginning with dead organic matter acted by microbes, then passing to detritivores and their consumers. For instance, dead bodies of small animals that were once part of the grazing food chain become incorporated in the detritus food chain as do the feces of grazing food animals. Functionally, the distinction between the two is of time lag between the direct consumption of living plants and ultimate utilisation of dead organic matter. The importance of the two food chains may differ in different ecosystems, in some grazing is more important, in others detritus is major pathway

The important point in Y-shaped model is that the two food chains are not isolated from each other. This Y- shaped model is more realistic and practical working model than the single-channel model because, (i) it confirms to stratified structure of ecosystems, (ii) it separates the grazing and detritus chains (direct consumption of living plants and utilization of dead organic matter respectively) in both time and space, and (iii) that the micro-consumers (absorptive bacteria, fungi) and the macro-consumers (phagotrophic animals) differ greatly size-metabolism relations. (E-P> Odum. 1983). It must however, be remembered that these models depict the basic pattern of energy flow in ecosystem. In practice, under natural conditions, the organisms are interrelated in a way that several food chains become interlocked results into a complex food web. We have already referred to food webs in grassland and in pond ecosystems. The complexity of food web depends on the length of the food chains. Thus in nature there operate multi-channel energy flows, but in these the channels belong to either of the two basic food chains i.e., will be either a grazing or a detritus food chain. Interlocking pattern of such several chains in food web of an ecosystem would lead to a multi-channel flow of energy. Thus in practice, under field conditions, we might face difficulties in measuring energetics of ecosystem.

1.9 BIOGEOCHEMICAL CYCLES

Nutrients, unlike energy are recycled in the ecosystem. There are about 40 chemical elements considered to be essential for living organisms. Materials are in limited quantity in the earth's system and to keep the system going continuously the only possibility is to regenerate the materials. The unique method evolved in nature is recycling materials continuously by linking them in cyclic changes. The macro-nutrients are C, H, 0, P, K, I, N, S, Mg, Ca, etc. which have cycles with atmosphere while micro-nutrients like Cu, Fe, Co etc are soil based form edaphic cycles. The bio-geo-chemical cycles are of two varieties-sedimentary cycles and gaseous cycles. In sedimentary cycles the main reservoir is the soil; the sedimentary and other types of rocks of earth's crust. The gaseous cycles have their main reservoir of nutrients in the atmosphere and oceans. Examples are the oxygen 'Cycle, carbon cycle, nitrogen cycle etc.

1.10 MATERIAL CYCLES

1.10.1 Nutrient Cycle

The supply of nutrients other than carbon dioxide, to an eco-system comes principally from the soil, but also to a smaller extent from the air, in rain and snow, and as dust. The supply of many nutrients is quite limited because they are in short supply in the soil and in other sources. Nutrients are cycled in such a way that they are both incorporated into plants and animals, or else are made available for plant uptake by the decomposition of dead plant and animal remains. The pathways from sources to sinks and back to sources, are termed elemental cycles, and they differ among the various elements. We consider briefly the three most important cycles, those of carbon, nitrogen and sulphur.

1.10.2 Carbon Cycle

Carbon is the basis of all organic molecules. It makes up our genetic material (DNA and RNA) and proteins, which are essential for life. Carbon is so special because of its ability to bond to almost any other molecule. The major element within our bodies is carbon.

The carbon cycle is the process through which carbon is cycled through the air, ground, plants, animals, and fossil fuels. Large amounts of carbon exist in the atmosphere as carbon dioxide (CO_2). Carbon dioxide is cycled by green plants during the process known as photosynthesis to make organic molecules (glucose, which is food). This is where the nourishment of every heterotrophic organism comes from. Animals do the opposite of plants—they release carbon dioxide back into the air as a waste product from respiration. (Note: Plants also undergo respiration to make food, but the majority of the carbon dioxide in the air comes from heterotrophic respiration). Decomposers, when they break down dead organic matter, release carbon dioxide into the air also. Decomposers are essential because without them, all of the carbon on the planet would eventually become locked up in dead carcasses and other trash. Decay permits carbon to be released back into the food web. Carbon is also stored in fossil fuels, such as coal, petroleum, and natural gas. When these are burned, carbon dioxide is also released back into the air. Volcanoes and fires also release large amounts of CO_2 into the atmosphere. Carbon dioxide can dissolve in water, where some of it is later returned back into the atmosphere. The rest can be taken to form calcium carbonate, which builds up shells, rocks, and skeletons of protozoans and coral.

The Carbon Cycle is a complex series of processes through which all of the carbon atoms in existence rotate. The same carbon atoms in your body today have been used in countless other molecules since time began. The wood burned just a few decades ago could have produced carbon dioxide which through photosynthesis became part of a plant. When you eat that plant, the same carbon from the wood which was burnt can become part of you. The carbon cycle is the great natural recycler of carbon atoms. Unfortunately, the extent of its importance is rarely stressed enough. Without the proper functioning of the carbon cycle, every aspect of life could be changed dramatically. Using energy from the sun, nature's carbon cycle goes around, from the atmosphere to the forest and back. Here is how it works. Trees absorb carbon dioxide from the air as they grow. In fact, about half their dry weight is this absorbed carbon. As old trees die and decay, or are consumed in a forest fire, their carbon is again released to the air as carbon dioxide. This is nature's carbon cycle.

When firewood is used as an energy source, part of the natural carbon cycle is brought into our homes to heat them. A fire on the hearth releases the solar energy stored by the tree as it grew. If the entire fuel cycle is considered, a clean burning fireplace will heat your home more efficiently and with lower environmental impact than any other fuel option. The other fuel options — oil, gas and coal — are fossil fuels, and when they are burned, old carbon that was buried deep within the earth is released to the atmosphere. The rising concentration of carbon dioxide from fossil fuel use is linked to global warming, climate change and the unusual weather we've seen in recent years.

A wood fire does not contribute to global warming because no more carbon dioxide is released than the natural forest would release if left untouched. Using wood for heat means less fossil fuels burned, less greenhouse gas emissions, and a healthier environment.

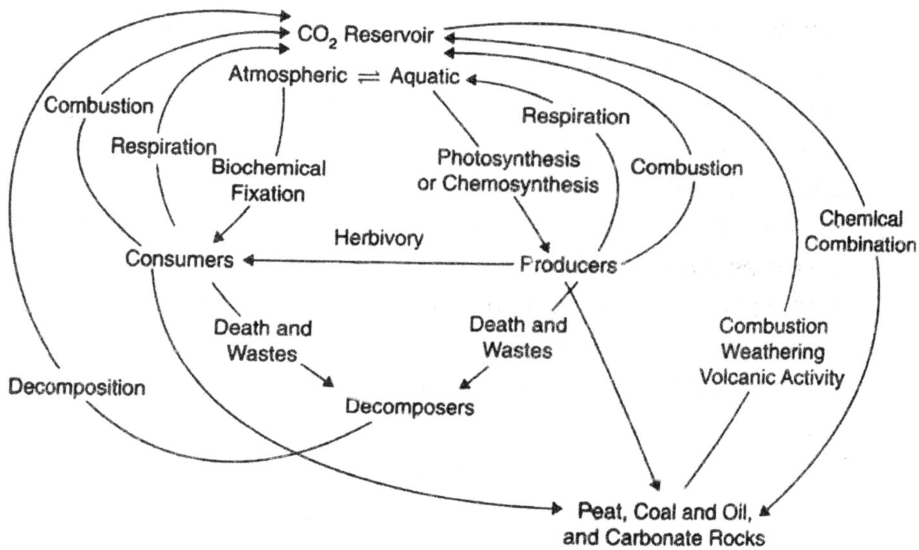

1.10.3 Nitrogen Cycle

Another important nutrient cycle is that of nitrogen. Nitrogen is a critically important element for all life. Proteins, which are constituents of all living cells, contain an average of 16% nitrogen by weight. Other complex nitrogenous substances important to life are nucleic acids and amino sugars. Without a continuous supply of nitrogen, life on earth would cease.

The nitrogen cycle is somewhat like the carbon cycle, but with a number of critical difference. Even though 79% of the earths atmosphere is composed of elemental nitrogen (N_2), this inert gas is entirely unavailable for uptake by most plants and animals. This is in stark contrast to the small amount of Carbon dioxide (0,03%) in the atmosphere, which is readily available for plant uptake. A relatively few microbes aro capable of fixing atmospheric nitrogen from the inorganic to the organic form. Such microbiological fixation averages 140 to 700 mg/m^2 year. In very fertile agricultural areas it can exceed 20000 mg/m^2 year.

A number of bacteria, fungi, and blue-green algae are known to be able to fix nitrogen. Nitrogen fixation involves the direct incorporation of atmospheric nitrogen into the organic body of the fixing organisms. The nitrogen fixers, constitute only a very small portion of these groups overall.

They can be divided into 1. symbiotic nitrogen fixers, which are largely bacteria, and which are associated with roots of legumes (members of pea and bean family) and some other flowering plants, and 2. free living nitrogen fixers. The genus Rhizobium includes those bacteria which in habit the nodules that develop on the roots of members of the pea and bean family. They are present in soil and infect the fine roots as seedlings grow. The roots produces a special nodule that houses the rhizobia, in which bacteria convert atmospheric nitrogen into the organic nitrogen constituents of their own cells. Since bacterial cells die very rapidly, this nitrogen becomes available to the higher plants. Crops of clover and beans actually add nitrogen to the soils in which they grown and eliminate the need for expensive fertilizers. A large scientific effort is under way in many countries to find bacteria that can form a similar association with the cereal grain crops.

The symbiotic nitrogen fixers seem to be confined to terrestrial ecosystems and have not been found in aquatic habitats, the one exception being a marine worm that attacks submerged wood. Among the non symbiotic nitrogen fixers are both aerobic and anaerobic free living bacteria as well as cyanobacteria. These occur in soils and in both marine and fresh waters and can add substantially to the nitrogen content of these environments. An additional but generally minor source of atmospheric nitrogen to soils and waters are lightning storms in which electrochemical nitrogen conversions take place.

Nitrogen enters the producer – consumer food chain when plants take it up from the soil solution either as nitrates or as the ammonium ion. Nitrate can also be converted to ammonia by denitrifying bacteria in the soil, especially by bacteria and fungi in water logged soils. Such conversion also occurs in low-oxygen conditions in lakes. The process is called denitrification. The nitrifying bacteria, in turn, can use ammonia nitrogen as a source of energy to synthesize their own protoplasm. This process occurs only slowly, if at all, under acidic conditions. First the ammonia is converted to nitrite by the bacterial genus Nitrosomonas, and the nitrite is then converted to nitrate by another genus, Nitrobacter. This two – step process is called nitrification. Both bacterial groups obtain their energy from this oxidation process and then utilize some of the energy to convert carbon dioxide to cellular carbon. Finally, after nitrate has been taken up and converted by higher plants and microbes into protein and nucleic acids, it is metabolized and returned to the major part of the cycle as waste products of that metabolism (inanimate organic nitrogen). Many heterotrophic bacteria and fungi in both soil and water utilize this organic nitrogen rich material, converting it and releasing it as inorganic ammonia in a process called amonificaton. Other parts of the cycle involves the release of gaseous nitrogen and nitric oxides, back into the atmosphere, although these are of limited significance

1.10.4 Sulfur Cycle

Sulfur is an important nutrient for organisms, being an key constituent of certain amino acids, proteins, and other biochemicals. Plants satisfy their nutritional needs for sulfur by assimilating simple mineral compounds from the environment. This mostly occurs as sulfate dissolved in soil water that is taken up by roots, or as gaseous sulfur dioxide that is absorbed by foliage in environments where the atmosphere is somewhat polluted with this gas. Animals obtain the sulfur they need by eating plants or other animals, and digesting and assimilating their organic forms of sulfur, which are then used to synthesize necessary sulfur-containing biochemicals.

In certain situations, particularly in intensively managed agriculture, the availability of biologically useful forms of sulfur can be a limiting factor to the productivity of plants, and application of a sulfate-containing fertilizer may prove to be beneficial. Sulfur compounds may also be associated with important environmental damages, as when sulfur dioxide damages vegetation, or when acidic drainages associated with sulfide minerals degrade ecosystems.

Sulfur (S) can occur in many chemical forms in the environment. These include organic and mineral forms, which can be chemically transformed by both biological and inorganic processes. Sulfur dioxide is a gas that can be toxic to plants at concentrations much smaller than one part per million in the atmosphere, and to animals at larger concentrations. There are many natural sources of emission of SO_2 to the atmosphere, such as volcanic eruptions and forest fires. Large emissions of SO_2 are also associated with human activities, especially the burning of coal and the processing of certain metal ores. In the atmosphere, SO_2 is oxidized to sulfate, an anion that occurs as a tiny particulate in which the negative charges are electrochemically balanced by the positive charges of cations, such as ammonium (NH + 4), calcium (Ca^{2+}), or hydrogen ion (H^+). These fine particulates can serve as condensation nuclei for the formation of ice crystals, which may settle from the atmosphere

The most important basic material of life is water, one of the jive panchabltutas. This is in limitted supply. Living forms on terrestrial areas are dependednt on water free from salts¬freshwater. Due to the solar heat water evaporates from the oceans and goes up as water vapour, and as it travel across on to land regions, gaining considerable height the vapour cools down to precipitate as water or snow. Of the total estimated water in the earth and its atmosphere

1.11 BIOMAGNIFICATION

Biomagnification, also known as bioamplification or biological magnification is the increase in concentration of a substance, such as the pesticide DDT, that occurs in a food chain as a consequence of:

- Persistence (Can't be broken down by environmental processes)
- Food chain energetics
- Low (or nonexistent) rate of internal degradation/excretion of the substance (often due to water-insolubility)

  Although sometimes used interchangeably with 'bioaccumulation,' an important distinction is drawn between the two, and with bioconcentration, it is also important to distinct between sustainable development and overexploitation in biomagnification.

- **Bioaccumulation** occurs *within* a trophic level, and is the increase in concentration of a substance in certain tissues of organisms bodies due to absorption from food and the environment.
- **Bioconcentration** is defined as occurring when uptake from the water is greater than excretion (Landrum and Fisher, 1999). Thus bioconcentration and bioaccumulation occur within an organism, and biomagnification occurs across trophic (food chain) levels.

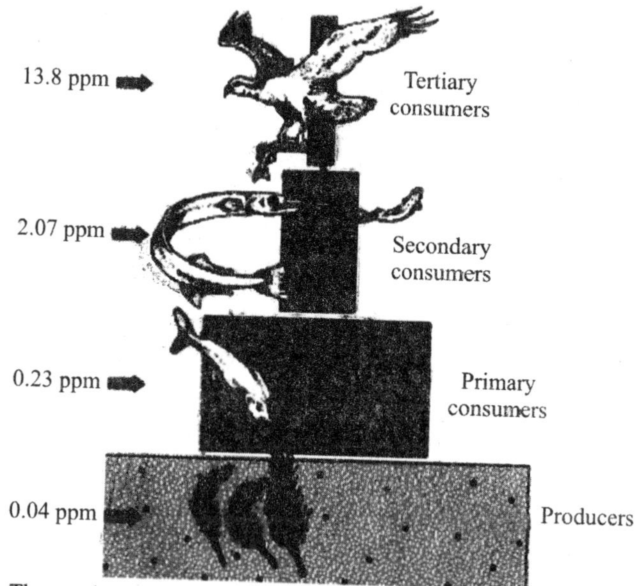

The numbers are representative values of the concentration in the tissues of DDT and its derivatives (in parts per million, ppm)

1.12 ECONOMIC VALUE

Ecosystem services are defined by Daily (1997) as "the conditions and processes through which natural ecosystems, and the species that make them up, sustain and fulfill human life." Humans benefit from the materials and processes provided by these services. Some of these materials such as seafood and lumber are clearly apparent while others are less conspicuous including waste assimilation, carbon cycling, storm buffering, and the hydrological cycle.

From a global perspective, ecosystems within the Basin contribute to the maintenance of the atmosphere by removing carbon dioxide and producing oxygen, by transforming nitrogenous wastes into less toxic forms, and by trapping pollutants. Global ecosystem services (Ecosystem services and functions) are large-scale processes that remain stable over long periods of time, thereby allowing biological systems to develop. Loss of these ecosystem services would result in significant changes in the stability of the global ecosystem, affecting all life on earth. However, the scale and complexity of the global ecosystem make it exceedingly difficult to detect changes; to attribute causes for the change; and to stop, or reverse, the processes leading to the change.

Key to understanding the importance of ecosystems and incorporating this in economic and other policy decision-making is to establish the link between a given ecosystem and its goods and services and how these are valued by individuals. The analysis is not about the absolute value of human life and the role of ecosystems in its provision, but about the marginal value of ecosystem functions, goods and services and tradeoffs between the competing uses (including conservation) of these.

Direct use value: Direct value involves human interaction with the ecosystem itself rather than via the services it provides. It may be consumptive or extractive use, such as fisheries and timber, or it may be non-consumptive, as with some recreational and educational activities.

Indirect use value: Indirect use value derives from services provided by the ecosystem. This might, for example, include the removal of nutrients, providing cleaner water to those downstream, the prevention of downstream flooding and diseases and provision of information.

Non-use value is associated with benefits derived simply from the knowledge that the ecosystem is maintained. By definition, it is not associated with any use of the resource or tangible benefit derived from it, although users of a resource might also attribute non-use value to it. It can be split into three basic components:

Existence value: Existence value derived simply from the satisfaction of knowing that ecosystems continue to exist, whether or not this might also benefit others (also associated with 'intrinsic value').

Bequest value: Associated with the knowledge that ecosystem and their services will be passed on to descendants to maintain the opportunity for them to enjoy it in the future.

Altruistic value: Derived from knowing that contemporaries can enjoy the goods and services ecosystems provide. Finally, another category not immediately associated with the initial distinction between use values and nonuse value includes:

Option value: An individual derives benefit from ensuring that ecosystem services will be available for his or her own *use in the future*. In this sense it is a form of use value, although it can be regarded as a form of insurance to provide for possible future use (often associated with the potential of genetic information inherent in biodiversity to be used for research, e.g. pharmaceuticals)

1.13 CARRYING CAPACITY

Carrying capacity in ecology refers to the level of land or resource use both by humans or animals, that can be sustained over the long term by the natural regenerative power of the environment. It is defined as the capacity of environment or an ecosystem to support and sustain a level of population under an ideal set of conditions. Carrying capacity is a ceiling of population, beyond which the environment fails to support the population leading to natural decrease in population. Usually when population growth reaches this phase - zero growth is achieved. Current growth and consumption patterns are placing increasing stress on ecosystems. Environmental degradation, biodiversity loss, deforestation, and the breakdown of social and economic systems are a few of the signs which indicate that ecosystems are stressed. Carrying capacity assumes that there are a finite number of people who can be supported without degrading the natural environment and social, economic and cultural systems and, as such, "is an indirect measure of the maximum level of stress that the ecosystem can maintain". (Barbier, Burgess and Folke 1994).

Carrying capacity affects all living things, including humans. It may be seen as a type of dynamic equilibrium. Carrying capacity is typically expressed as the number of animals of a given type which can be supported in a given area. Carrying capacity is also interpreted more broadly as the number of living things — plants as well as animals — any area of land or water can support at any one time. Different life forms will have different carrying capacity within the same area. Carrying capacity is usually limited by some aspect of a species' habitat requirements. These requirements include the quantity and quality of available food, shelter, water, and space, and the suitability of their arrangement. Carrying capacity can fluctuate over time. For example, changes in the local environment during seasonal change can affect carrying capacity. Alteration of habitat quality or quantity may increase or decrease carrying capacity. Environmental degradation may reduce it for affected species. Activities such as development or pollution tend not to be aimed at intentionally reducing carrying capacity but often have this impact. Intentional intervention may be based on a particular management philosophy or practice. Management of an area of land or water in relation to its carrying capacity for certain species can be subject to question and controversy. Whether and/or how it is appropriate for humans to intervene in natural systems is often a part of such questions and controversy. Management may be defined as intentional choice based on human goals and objectives. Such goals and objectives are open to question by other groups and individuals.

REVIEW QUESTIONS

1. Explain concept of ecology.
2. Discuss the structural and functional concepts.
3. Write a notes on components of an ecosystem.
4. Explain the food chain in ecosystem
5. Write short notes on
   (i) Food webs (ii) Ecological pyramids
6. How do you explain the energy flow in an ecological system?
7. Write notes on
   (i) Single channel energy models (ii) γ-shaped energy model
8. Write notes on
   (i) Nutrient cycle (ii) Carbon cycle
   (iii) Nitrogen cycle (iv) Sulfur cycle
9. Write an essay on Biomagnification.
10. Define carrying capacity and explain it briefly.

CHAPTER 2

NATURAL RESOURCES

2.0 CLASSIFICATION OF RESOURCES

All the resources are classified based on quantity, mutability and reusability, men's activity, quality, availability. Fig. 2.1 shows the classification of resources. In this chapter, the classification related to living

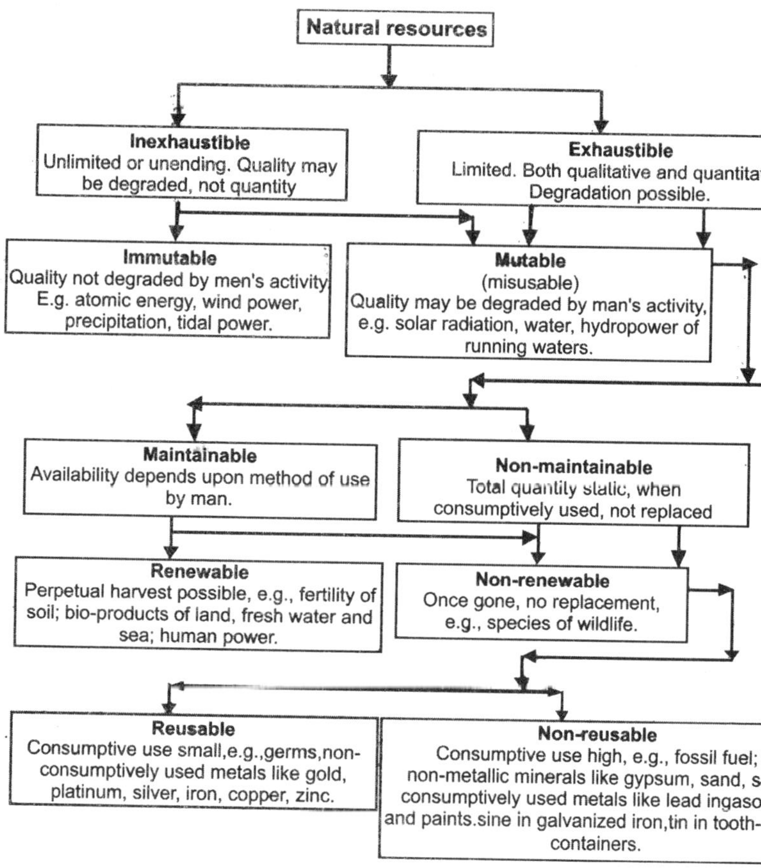

Fig. 2.1 Classification of Natural Resources

and nonliving resources and renewable and non renewable resources are presented.

2.1 LIVING AND NON-LIVING RESOURCES

Resources are also classified as biotic (or living) and a biotic (or non-living) resources. Forest, agriculture, fish and wildlife falling in the first category, and land, water minerals and etc., in the second category. Based on quantity, mutability and reusability all resources are classified as inexhaustible and exhaustible these are the main categories of resources based on their stocks in nature. Inexhaustible resources are further divided into two groups-immutable and mutable resources, depending upon the possibility of their qualitative degradation as a result of man's activity. On the other hand, exhaustible resources, being limited in occurrence, are vulnerable to both quantitative and qualitative degradation. Thus, their availability depends upon the method of use, so these care classed into maintainable and non-maintainable resources. Supply of maintainable resources could be made to last for long through wise use, however, some of them have capacity to reproduce-called non-renewable resources. Non-maintainable resources have static supply and when destroyed or consumptively used are irreplaceable, yet some of them have potentiality for reuse-called reusable resources, and rest whose permanent exhaustion is certain are called non-reusable resources. It is convenient to divide resources into three general categories: 1. renewable resources, 2. non-renewable resources that can be recycled, and 3. non-renewable resources that cannot be cycled.

2.2 RENEWABLE AND NON-RENEWABLE RESOURCES

2.2.1 Renewable Resources

One cause of environmental degradation is the overuse of common-property resources, which are owned by none and available to all users free of charge. Most are potentially renewable. The term global commons refers to portions of Earth and its surrounding space that lie beyond the territorial claims of any nation. Examples include clean air, the open ocean, fish in the open ocean, migratory birds, Antarctica, gases of the lower atmosphere, the ozone content of the stratosphere, and space, including the vast high seas, as much as 70% of Earth's surface is commons.

2.2.2 Recyclable Non-renewable Resources

All non-energy mineral resources which occur in the Earth's crust like ores of copper, aluminum, mercury, and other metals; deposits of fertilizer nutrients such as phosphate rock and potassium; and minerals that are used in their natural state such as asbestos, clay, and mica are considered as non renewable resources which can be recycled. As these deposits are mined, they are not replaced, and hence such materials are considered non-renewable. However, it is possible, at least in theory, for people to collect these materials or elements they used and recycle them. Recycling involves collecting and reprocessing a resource into new products. For example, glass bottles can be crushed and melted to make new bottles or other glass items. Reuse involves using a resource over and over in the same form. In practice, we never completely exhaust a nonrenewable mineral resource. However, a mineral resource becomes economically depleted when the costs of finding, extracting, transporting, and processing exceeds the amount earned from them. At that point we have five choices: recycle or reuse existing supplies, waste less, use less, try to develop a substitute, or do without and wait millions of years for more to be produced.

2.2.3 Non-Recyclable – Non-renewable Resources

Resources that exist in a fixed quantity in Earth's crust and thus theoretically can be completely used up are called non-renewable (or exhaustible) resources. On a time scale of millions to billions of years, such resources can be renewed by geological processes. However, on the much shorter human time scale of hundreds to thousands of years, these resources can be depleted much faster than they are formed. Fossil fuels are derived from organic matter that accumulated during hundreds of millions of years of early biogeological history. There is no way of recycling the energy in fossil fuels.

Resources that cannot be recycled are those "mineral" energy resources, namely, fossil fuels (coal, oil, and natural gas), that presently supply better than 90 percent of our energy, and uranium that is used for nuclear (atomic) power. Non-renewable energy resources, such as coal, oil, and natural gas, can't be recycled or reused. Once burned, the useful energy in these fossil fuels is gone, leaving behind waste "heat and polluting exhaust gases. Most of the per capita economic growth has been fueled by relatively cheap non-renewable oil, which is expected to be economically depleted within 40 to 80 years.

2.3 WATER RESOURCES

Water is the most essential commodity for all living beings for their survival. Human beings, animals, aquaculture, wildlife, birds, insects & flies, trees, plants, crops, microorganisms and the like consume large quantities of water for their growth and other needs. It is said that even the human body is full of water that is more than 60% of its volume is of water. Most of the ancient cities were observed to have developed on the banks of major rivers. This also amply signifies the role of water in the development of civilization even in pre-historic days. So far, people were treating water as only the life-sustaining commodity. But with the rapid growth of industries developed for creating the infrastructure facilities for various needs and for rapid growth in urbanization, the demand of water is increasing day by day. However, water is a finite source, i.e. in hydrosphere, roughly constant amount of water is in circulation at any one time but much less than 1 percent actually available for human use. It has been estimated that about 1400 X 10 15 M3 of water exist on the earth, of this 97% is in oceans as saline water, 2% in frozen in ice caps. The fresh water availability in lakes, streams and rivers and underground form much less than 1% of the total water resources of the earth. Therefore, there is no alternative but to make most efficient utilization of the available water. The idea about the availability of water from various sources can be obtained from the knowledge of hydrology.

2.4 USE AND OVER UTILIZATION OF SURFACE AND GROUND WATER

2.4.1 Agriculture Needs of Water for the Crops

As our country is essentially an agricultural based country, the crops are to be developed for the production of different types of food grains. The requirement of water varies from crop to crop. Different research stations are busy in identifying the water needs of all the crops. Most of these crops are shallow rooted, thus water being extracted from top layers of the soil. Soil moisture available in the top layers is essential for such crops. Hence, it is advocated to keep the topsoil always moist so that crops do not wilt under no water condition. Several scientific methods of ploughing are developed to maintain the moisture in the topsoil for longer periods. Requirement of maintaining the water storage for breeding the fish and prawn culture is attaining importance whenever we go for storage project. Since fish cannot travel against flow when the velocity is high, special structures are provided adjacent to the weirs to maintain low velocities. Such structures are known as fish ladders.

The different purposes of the multipurpose projects have their conflicting affect on the environment. However, depending on the priority of the water needs of the area to be served, the planning has to proceed in such a way that the impact on environment is kept to be minimum while serving the urgent and primary needs of the people of the area.

2.4.2 Groundwater and Aquifers

Water that is available in the deeper layers of the earth is known as Groundwater. This water has been trapped inside the earth's crust for several centuries. The water that is lying under the ground has the capacity to move in general direction of its slope with a very small velocity. The bodies that contain such water are known as Aquifers. Water from these aquifers can be drawn by digging wells and pumping water from these wells. Some of these wells can supply very large quantities of water. In some cases, the drinking water needs of cities are completely met by the groundwater bodies. In the agriculture sector also, groundwater is supporting the growth of crops enormously in certain agricultural dominated areas.

2.4.3 Groundwater and Wells

Wells are used to bring groundwater to the land surface by means of pumps. Wells can be deep wells and shallow wells depending upon the depth at which ground water is available. Sometimes open dug wells are used where the water table is high. In the case of deeper and hard rock aquifers, Tube wells are constructed. In such cases deep well pumps such as turbine pumps or jet pumps are used to lift water to the surface of the ground. These tube wells in many cases, yield significantly high discharges. In a way, groundwater is the water that is available at the place of its use. Thus this water can be lifted from the wells in the agricultural fields and supplied to the crops raised in that field. Hence, there is no need to send water to long distances. This incidentally reduces the water losses considerably. Use of ground water for nearby areas also does not pose the problems of environmental degradation.

2.4.4 Floods and Droughts

2.4.4.1 *Floods*

The flood hazard itself cannot be prevented, but thorough understanding of the land conditions which are prone to a given hazard and the processes which could culminate in the damage to life and property, it is possible to minimize the damage through preparedness for a particular eventuality. Flooding takes place when the river channels are unable to contain the discharge. In the tropical countries, floods are caused by various factors: (i) Climatologically (rain), (ii) part climatologically (coastal storm surges, estuarine interactions between stream flow and tidal conditions) and (iii) others (failure of dams and other control works, excessive release from dams). Floods could get intensified because of basin characteristics, network characteristics, and channel characteristics, each of which has both stable (unvarying) and variable components (Table 2.1) unvarying and variable characteristics.

Table 2.1

	Stable Unvarying	**Variable**
Basin	Slope, attitude, shape, et., of the basin	Arising from the interactions between climate, geology, soil type, vegetation cover, etc., which are manifested through the storage capacity of soil and bedrock, extent of infiltration, and transmissibility of soil and bedrock, that are affected by anthropogenic activities.
Network Characteristics	Patters	Surface storage, under-drainage, channel length, contributing or source area etc.
Channel Characteristics	Slope, flood control and river regulation works	Roughness, load, shape, storage

The improper land-use practices accentuated the flood devastation. There are hardly any forests left in the catchment area of the rivers. It is well known that the forest areas are characterized by high infiltration capacity and transmissibility. The infiltration capacity of the forest areas is 2-3 times greater than in the open fields. The surface runoff in the forested areas may be as little as one-tenth of that of the open fields. There is hardly any protective vegetation on the banks of the Save and Limpopo rivers. Crops are grown right to the edges of the rivers, even on river slopes. Thus, the combination of absence of forest cover on one hand, and inappropriate farming practices on the other, intensified the floods. Floods can be mitigated by structural, water control and non-structural measures. The structural methods include dams, reservoirs, and retarding basins, channel management and embarkments. The water control methods include flood proofing and catchment modifications. Schemes of drainage and flood protection, flood forecasting, flood warning and emergency preparedness systems, flood insurance, public information and education, and flood relief constitute the non-structural methods.

2.4.4.2 *Drought*

Drought is lack or insufficiency of rain for an extended period that causes considerable hydrologic imbalances and consequently water shortages, stream flow reductions and depletion of groundwater levels and soil moisture. Drought is the most serious physical hazard to agriculture in nearly every part of the world. Drought not only leads to serious economic consequences but also leaves behind untold human misery. Among all the natural disasters, drought affects largest number of people in the world. Shortage of water for even the basic needs is the main problem in the drought areas. Even the shallow rooted crops do not grow in such areas. Getting sufficient drinking water is another problem needing immediate attention in these areas. Some measures like infiltration wells, underground dams, small watersheds, are being taken up to alleviate the sufferings of the people residing in the drought prone areas. Certain advance techniques such as Cloud Seeding and Artificial Rains are also being tried with varying successes. However, these methods are quit expensive and unpredictable in their success. Scant rains for extensive periods also lead to ecological changes. Ultimately, Government has found reasonable remedies in the form of development of small watersheds in such areas.

2.4.5 Dams: Benefits and Problems

Dams are the major structures in any reservoir from the point of view of structural importance; design details and cost. The dams are of different types depending on different criteria. Depending on the material used for construction, dams can be: masonry dams, concrete dams, Earthen dams, rock fill dams, steel dams and timber dams. Based on the design, the dams can be: gravity dams, arch dams, buttress dams, and multiple arch dams. Similarly, based on the purpose, the dams are known as overflow dams and non-overflow dams. The masonry and concrete dams are more or less leak proof and hence seepage is not possible. However in the case of earthen and rock fill dams, seepage of water can be expected, in good quantities and therefore, possibilities of water logging on downstream side will be the adverse effect on the environment.

The water resources projects are constructed to several purposes depending on the needs of people of the area to be served. Whenever the projects are developed to supply water for different purpose, the projects are termed as multi-purpose projects. The different purposes can be: irrigation and agriculture, hydropower generation, drinking water supply, water for Industries, flood control, navigation, recreation and amusement parks and afforestaion. Of all the above purposes, irrigation and agriculture occupies higher priority, as the production of necessary food grains for one billion population of the country is the primary concern to us.

In order to develop the industries and other power needs, the next priority is the Hydropower development. The emphasis is increasing on the hydropower, as the natural resources for other forms of energy such as thermal, are becoming scarce. Due to rapid development of urban areas, scarcity of drinking water has surfaced in most of the cities. Hence, the present emphasis is on bringing water to the cities from storage reservoirs. Slowly, water needs for drinking purposes is occupying priority when compared to other needs. In the flood plains, the problem of inundation of adjacent habitated areas is a priority, as during every flood the losses in respect of human and cattle life, crops, property and fertile soils, are bringing misery to the people. With every increasing movement of men and material, the transportation by navigation is also recognized as a viable mode. In the recent times, water has been used for recreational purposes also. In some countries, water sports are gaining popularity. In order to develop greenery in many dry areas, Government is encouraging people to go in for tree plantations on a large scale resulting in afforestation.

2.5 MINERAL RESOURCES

In any country, the growth and development of industry depends to a very great extent, on the availability, quality and reserves of deposits of rocks and minerals of economic importance. India, like other countries of the world, is not self sufficient in all mineral requirements. Every year, millions of foreign exchange is spent for importing necessary metals and minerals. At the same time, the country occupies very prominent position in the world market for the production of some other minerals, chief among them are manganese, mica, iron, aluminum, magnesium and titanium besides high grade refractories and thorium minerals. The country is meeting the requirements by importing nonferrous metals like copper, zinc and lead. Extensive prospecting work is being done for the ore-minerals of these and other metals, and the indications are quite encouraging. It is also deficient in sulphur and petroleum and is mostly dependent on other countries for them. Mineral resources can be classified under three main types. They are: metallic minerals, non-metallic minerals (including mineral fuels) and atomic minerals.

2.5.1 Use and Exploitation

Mineral resources exploitation took place at 621 exclusive mineral deposits of which 381 deposits were with exploitation site regarded as protected deposit area. The statistics shows that 182 exclusive mineral deposits were only with protected deposit sites being approved and 58 exclusive deposits were undergoing approval procedure in 1998. Besides exclusive deposit exploitation was in progress also at 242 non-limited mineral deposits.In1998, stabilisation trend in mining prevailed (brown coal and lignite, magnetite and salt) when compared with the previous period. In some cases a slight increase was recorded as brick clay, high-content limestone, lime stones and raw materials for cement production.

From the total amount of coal mined, approximately 85% was used for energy production, 15% was distributed in a form of graded coal. In 1998, from the total amount of 46,312 tonnes of crude oil exploited 4 166 tonnes were of non-paraffin oil and 42 146 tonnes were of semi-paraffin oil. Gajary and Dubrava crude oil deposits had the largest share in crude oil exploitation. Their contribution to crude oil exploitation was 77% of the total exploitation.

2.5.2 Environmental Effects

Mining of minerals, being an environmentally unfriendly activity, has attracted attention from the stand point of environmental impacts and their mitigation. Another fact about mining is that it is a site specific activity and is only an intermediate use of the land because mining is done at the sites where the minerals exist and the land is of no use to the mining companies in the pre and post mining times. Mining affects all the components of environment and the impacts are permanent/temporary, beneficial/harmful, repairable/irreparable, and reversible/ irreversible. The impacts of various mining and associated activities on the environmental components are discussed briefly in the following paragraphs.

2.5.2.1 *Ecological Impacts of Opencast Mining*

- Removal of all vegetation (flora) and thereby fauna from the area required for mining and other purposes
- Pollution of water in the surrounding water bodies due to leaching from overburden dumps and due to the pollutants from the other activities. This affects the aquatic ecology of these water bodies.
- Dust in atmosphere, contributed by mining and associated activities, when deposited on the leaves of the plants in the surrounding areas may retard their growth.
- Noise and vibrations due to blasting and operation of the machines drive away the wild animals and birds from the nearby forests.
- Water scarcity caused due to the impacts of opencast mining on water regime affects the growth of vegetation and agriculture in and around the complexes.

2.5.2.2 *Ecological Impacts of Underground Mining*

- Clearing of area for developing shaft/incline complex, infrastructure, colonies, etc. may require removal of some vegetation and thereby driving away the fauna.
- Water scarcity, caused due to the impacts of mining on water regime, along with pumping and release of polluted water on the surface may affect vegetation in the surrounding areas.
- Top-soil in tensile zones of subsiding areas may loose its vegetation supporting capability.
- Release of polluted water from the underground mines into the surface water bodies may affect their aquatic ecology.

2.5.2.3 *Ecological Impacts of mineral handling and preparation*

- Land clearance of almost all vegetation in the area earmarked for the construction of the mineral handling and preparation units.
- Disturbances to fauna of the nearby areas from the noise and vibrations from the mineral handling and preparation units.
- Impacts on aquatic ecology due to discharge of effluents from the units.
- Retardation in vegetation growth in neighboring areas due to deposition of dust on the leafs.

2.5.2.4 *Ecological impacts of other activities*

- The growth of mining complexes need land and thus affects the ecology of the land and the surrounding areas.
- Cutting and felling of the trees to meet the timber requirement for various purposes.
- Other impacts are similar to those of the activities mentioned above.

2.5.2.5 *Action to Minimize the Impacts*

It is evident that mining and associated activities have considerable impacts on the ecology of the mining and the surrounding areas. These impacts are evident in most of the mining complexes in the country. In order to minimize the impacts the following actions can be thought out.

- Plan the mining layout so as to have the least requirement of the forest land and take necessary steps for reclamation of the mined out land so that the forest land taken for the mining purposes can be brought back to forest use.

- Develop a suitable compensatory forest.
- Cut the trees to the minimum possible extent and to preserve the flora it would be appropriate to uproot the trees and plants and then establish them at suitable locations, may be in the areas ermarked for compensatory afforestation.
- Develop a *flora bank* to preserve the typical floral species of the area so that these can be replanted and developed as and when needed.
- Surface layout of the mining complexes be designed to have the least impacts on the ecology of the area.
- The noise and vibration producing activities in the mines and the associated activities be planned to have the minimum possible intensity and impact on the wild life in the surrounding area.

2.5.2.6 *Impacts on Environment*

The mining and associated activities in the mining complexes not only pollute the air but also cause noise pollution. These are briefly given below.

Impacts of opencast mining

- Removal of vegetation from the area designated for mining and other purposes produces dust which when air-borne causes an increase in the concentration of SPM in the surrounding air.
- Removal, handling, transportation and storage of soils also causes an increase in the concentration of SPM in the atmosphere. The use of diesel equipment in these activities causes an increase in the level of NO_x.
- Drilling and blasting of overburden and the mineral contribute SPM and explosive fumes into the atmosphere.
- In-pit crushing, loading and transportation of the mineral and the overburden rock mass and making the dumps contribute SPM and NO_x.
- Minerals and rock mass having sulfur and its compounds may contribute SO_2.
- Making of the overburden dumps and the use of diesel equipment for this purpose contribute SPM and NO_x.
- Some of the sedimentary rocks may have CH_4 and when mined they may contribute this gas to the surrounding air.

- Fires in opencast mines contribute heat, SPM, SO_2, CO_2 and CO.
- Use of petrol vehicles in the mines contributes hydrocarbons and lead.
- The equipment used in the opencast mines for various purposes including the transport of the overburden and mineral generate continuous noise, while blasting produces impulsive noise. All the noise generated in the mine does not become ambient nose as the noise generated from the different sources gets reflected and refracted and ultimately the resultant of all the noises after reflection and refraction reaches the surface to become the ambient noise.

Impacts of underground mining

- The exhaust air from the underground mines contributes SPM, CO_2, CH_4, NO_x, SO_2, and other pollutants.
- The surface activities in the underground mining areas, e.g., diesel generating stations, boilers, etc. contribute SPM, NO_x, and CO_2.
- The machines and equipment installed at the shafts, inclines, compressor houses, workshops, etc. generate noise, which tends to become ambient noise as it is generated on the surface.

Impacts of mineral preparation activities

- Mineral handling, mineral preparation and associated activities mainly contribute SPM to the atmosphere. In the mineral preparation plants having chemical processes producing gases the atmosphere may get polluted due to emission of the gases.
- The crushers, conveyors and other equipment installed in the mineral handling and preparation plants produce continuous noise.

Impacts of other activities

- Domestic burning of coal and other fuels, and open burning of coal produce SPM and CO_2 and contribute them to the atmosphere.
- The transport network using diesel and petrol vehicles and generation of power by DG sets, and other activities in the mining complexes contribute SPM, NO_x, CO_2, CO and other atmospheric pollutants.
- Smoking contributes SPM and CO_2.
- Industrial activities depending upon their processes, inputs and outputs contribute various air pollutants.
- Paddy cultivation contributes CH_4.

- Use of pesticides, insecticides and other chemicals for spraying and in agricultural and horticultural activities may contribute air borne chemicals.
- The constructional activities contribute SPM to the atmosphere.
- Domestic activities, playing of music and loud speakers, plying of vehicles on the roads, DG sets, etc. produce noise which becomes ambient noise.

2.5.2.7 *Remedial Measures*

It is noted from the above mentioned impacts of the different activities on the atmosphere that the mining and associated activities not only contribute to the ambient air pollution but also to the ambient noise situation. Atmospheric pollution due to the mining and associated activities can be minimized by planning the activities in such a manner that the generation of the pollutants is minimum possible. In addition provisions may be made for arresting the dust by making suitable green belts. Some of the measures are as outlined below.

- Generation of dust in the removal of the vegetation and soils can be minimized by maintaining adequate moisture in the soil. This can also be expected to improve the efficiency of these operations as in dusty atmosphere the efficiency decreases.
- Use of dust extractors with the drill machines can be expected to minimize air pollution due to drilling.
- By optimizing the blast design the generation of dust due to blasting can be reduced.
- Proper maintenance of the haul roads can minimize the generation of air borne dust due to movement of dumpers on them.
- Water spraying at the transfer points tends to reduce air pollution.
- Enclosing the mineral handling and preparation units tend to reduce the contribution of SPM to the atmosphere.
- Proper maintenance of the equipment and machines in the mines and other places in the complexes helps not only in minimizing the contributions to the air pollution but also the noise generation.
- In the residential locations playing of the loud speakers specially after 9.00 pm should be avoided to minimize the noise levels.
- The locations of the residential locations should be planned such that they are on the up wind side of the mines and plants so that for most of times the atmospheric pollutants are taken away rather then being brought towards these locations.

- Green belts of adequate widths, say 25 -50 m, may be planned between the residential areas and the mines not only to attenuate noise but also to arrest dust.

2.6 LAND RESOURCES

Land use refers to man's activities on land which are directly related to the land. Land cover, on the other hand, describes the vegetation and artificial constructions covering the land surface. The types of land use and land cover categorization developed in any classification system can be related to systems for classifying land capability, vulnerability to certain management practices, and potential for any particular activity or land value, either intrinsic or speculative. Concepts concerning land cover and land use activity are closely related and in many cases have been used interchangeably. Land use planning creates the prerequisite required to achieve a type of land use, which is sustainable, socially and environmentally compatible, socially desirable and economically sound. Land use planning is not a standardized procedure which is uniform in its application. World wide its content is based on an initial regional or local situation analysis. The uses of land use / land cover maps will be utilized as a basic data which provides the foundation for allocating new potential users. Land use / Land cover (LU / LC) maps can be used for over all development of the region. And the land use map will serve as a base in the overall planning of agriculture, industrial development and environmental impacts and interactions within the region. Land use /land cover classification system has been primarily developed for use with remote sensing data in order to meet the required basic information on land use / land cover for agro-climatic zones planning.

Land is the critical resource for the sustenance of living organisms. The ever growing population is exerting pressure on the finite resources. Hence it has become a necessity to know about the present or existing use of land so as to plan for its sustained optimal use. In our country efforts are being made for a rational and scientific regional and urban planning. Reliable and integrated information, which is necessary for land use planning, is not available. The present status of availability of information on LU/LC in the form of thematic layers and collateral data are insufficient and so cannot be provided latest information. For the last two decades Satellite Remote Sensing techniques were used for the generation of Land Use information. Land is the most important natural endowment on which all the man's activities are based. Growing population and increased human activities are exerting pressures on limited land resources. This is evident by the decrease in per-capita available cultivable land from 0.48 ha. in 1951 to 0.22 ha. in 1991. Land use refers to man's activities on land which are directly related to the land. Land cover, on the other hand, describes the vegetation and artificial constructions covering the land surface. The types of land use and land cover categorization developed in any classification system can be related to systems for classifying land capability, vulnerability to certain management practices, and potential for any particular activity or land value, either intrinsic or speculative. Concepts concerning land cover and land use activity are closely related and in many cases have been used interchangeably. Land use and land cover have a direct relationship with environmental characteristics and processes, including the productivity of the land, species diversity, climate, biogeochemistry and the hydrologic cycle.

The characteristics of land cover and land use have an impact on the climate, biogeochemistry, hydrology, and the diversity and abundance of terrestrial species. Large pressure of growing population increased demand for food, fodder and fuel wood combined with industrial activities have essentially led to rapid change in land use / land cover patterns particularly in developing countries. The ever growing population and limited availability of land necessitates proper utilization of available land through scientific land use planning. Data on land use / land cover patterns, their spatial distribution and changes are the prerequisites for making development plans. Digital classification techniques through stratification approach, with limited ground truth, could be used for mapping broad categories of forests and their density classes with aerial extent and location. Information with respect to forest cover, their types, classes, density, quality and distribution, is of prime need for adopting conservation measures, eco-balance and planning-cum management aspect. GIS and Remote Sensing are gradually becoming an integrated technology that is being widely used in various applications. These two technologies are complementary, as they are simply variants of the digital spatial data. The spatial display techniques for GIS and Remote Sensing information has greatly advanced within the past two decades. Technological advances in image processing and visualization techniques have developed display and interpretation mechanisms for the analysis of all forms and sources of geographical information. Recent development trends have shown that automated and computer assisted techniques for the interpretation and change detection is becoming widely used. Remote Sensing has been recognized as a useful means of supplying timely and up-to-date information on activities within the urban environment. It has been noted that the integration of GIS technology with Remote Sensing will provide the optimum information context and analysis capabilities and thus be of benefit to urban land-use planners.

Land use change may involve either a shift to a different use, such as agricultural land changing to residential, or an intensification of the existing use, such as from light to heavy industry. The dual role of human activity in contributing to the causes and experiencing the effects of global change processes brought about by land use changes emphasizes the need for better understanding of the interaction between humans and the environment. This need becomes more imperative as changes in land use become more rapid. Knowledge of land use and land cover is important for many planning and management activities and is considered as an essential element for modeling and understanding the earth as a system. Land use is in part determined by environmental factors such as soil characteristics, climate, topography, and vegetation. Large-scale approaches often utilize a watershed approach and include the use of GIS to evaluate land use, soil type, elevation, and other mappable characteristics to explain the spatial distribution of non point source contamination in groundwater and surface water.

2.6.1 Land as a Resource

Soil is a major component of land system, which provides a medium for plant growth. The potentials for sustainable use under agriculture, horticulture, silvipasture and forestry as well as its response to irrigation and other management practices are controlled by its inherent qualities and characteristics. Earlier, soil surveys have been carried out using topographic maps and cadastral maps as database. The development of aerial photo interpretation technique in late sixties in India substantially augmented the efficacy of soil mapping programme. Application of remote sensing for soil mapping helps to overcome these shortcomings encountered in traditional system. Soil resources information need to be translated into interpretative exits, which are indicative of the potentials and limitations of land for sustainable agriculture development, to generate action plan. Land capability, land irrigablility and land suitability are some of the derivative information commonly used.

2.6.2 Land Capability Classification

It is an interpretative grouping of soils based on 1. Inherent soil characteristics 2. External land features 3. Environmental factors. In the land capability classification there are '8' classes. Classes I, II, III include the land suited for regular cultivation. Class IV land is fairly good for cultivation but its safe use is limited by natural features such as slope, erosion, unfavorable soil characteristics and adverse climate. Classes V, VI, and VII are not suited for any cultivation but may be used for grazing or forestry according to quality of the land. Class VIII land is suited only for wild life recreation. The land capability classification is based on "AISLUS" (All India Soil and Land Use Survey) guidelines.

2.6.3 Land Degradation

Land degradation is a concept in which the value of the biophysical environment is affected by one or more combination of human-induced processes acting upon the land. Natural hazards are excluded as a cause; however human activities can indirectly affect phenomena such as floods and bushfires. It is estimated that up to 40% of the world's agricultural land is seriously degraded. Land degradation refers to a decline in the overall quality of soil, water or vegetation condition commonly caused by human activities. Degradation is also considered to include a change in the ground cover to less palatable species, or a change from predominantly perennial grasses to predominantly annual grasses. Environmental dilapidation is brought about by pollution especially in urban areas, which not only experience a rapid growth of population due to high fertility rates, low mortality and increasing rural-urban migration, but also due to the rapid industrialization. Major ecological and socio-economic crisis are perpetrated by land/soil degradation. Direct impacts of agricultural development on the environment arise from farming activities, which contribute to soil erosion, salinity/ brackishness of land and loss of nutrients. The Green Revolution has been accompanied by over exploitation of land and water resources and use of fertilizers and pesticides have increased manifold. In the race to urbanize virgin territory, there has been random violation of the land laws. The current trend of economic and industrial development coupled with the steady growth of human as well as livestock population have been the major reasons behind the incidence of land gradation in India. These factors exert pressure on limited land resources of the country for agricultural, industrial and housing needs of the growing population. It is the lands under cultivation which face the biggest challenge of land degradation in India. Go through India housing to know all about Land Degradation.

The major causes include: land clearance, such as clear cutting and deforestation, agricultural depletion of soil nutrients through poor farming practices, livestock including overgrazing, irrigation and over drafting, urban sprawl and commercial development, land pollution including industrial waste, vehicle off-roading, and quarrying of stone, sand, ore and minerals

The main outcome of land degradation is a substantial reduction in the productivity of the land. The major stresses on vulnerable land include: Accelerated soil erosion by wind and water, soil acidification or alkalinisation, salination, destruction of soil structure including loss of organic matter and derelict soil.

Over cutting of vegetation occurs when people cut forests, woodlands and scrublands--to obtain timber, fuel wood and other products--at a pace exceeding the rate of natural regrowth. This is frequent in semi-arid environments, where fuelwood shortages are often severe.

Agricultural activities that can cause land degradation include shifting cultivation without adequate fallow periods, absence of soil conservation measures, and cultivation of fragile or marginal lands, unbalanced fertilizer use, and a host of possible problems arising from faulty planning or management of irrigation. They are a major factor in Sri Lanka and the dominant one in Bangladesh.

We often assume that land degradation only affects soil fertility. However, the effects of land degradation often more significantly affect receiving water courses (rivers, wetlands and lakes) since soil, along with nutrients and contaminants associated with soil, are delivered in large quantities to environments that respond detrimentally to their input. Land degradation therefore has potentially disastrous effects on lakes and reservoirs that are designed to alleviate flooding, provide irrigation, and generate hydroelectricity.

2.6.4 Man Induced Landslides

The term "landslide" is used to describe a wide variety of processes that result in the perceptible downward and outward movement of soil, rock, and vegetation under gravitational influence. The materials may move by: falling, toppling, sliding, spreading, or flowing. Although landslides are primarily associated with steep slopes, they also can occur in areas of generally low relief. In these areas landslides occur as cut-and-fill failures like highway and building excavations, river bluff failures, lateral spreading landslides, the collapse of mine-waste piles especially coal, and a wide variety of slope failures associated with quarries and open-pit mines.

Landslides can be triggered by both natural and man-induced changes in the environment. The geologic history of an area, as well as activities associated with human occupation, directly determines, or contributes to the conditions that lead to slope failure. The basic causes of slope instability are fairly well known. Land slides can be explained into three ways. They can be inherent, such as weaknesses in the composition or structure of the rock. They can be variable such as heavy and such as seismic or volcanic activity; or due to new environmental conditions.

Human activities triggering landslides are mainly associated with construction and involve changes in slope and in surface-water and ground-water regimes. Changes in slope result from terracing for agriculture, cut-and-fill construction for highways, the construction –of buildings and railroads, and mining operations. If these activities and facilities are ill-conceived, or improperly designed or constructed, they can increase slope angle, decrease toe or lateral support, or load the head of an existing or potential landslide. Changes in irrigation or surface runoff can cause changes in surface drainage and can increase erosion or contribute to loading a slope or raising the ground-water table.

There are a number of natural factors that can cause slope failure. Some of these, such as long-term or cyclic climate changes, are not discernible without instrumentation and/or long-term record-keeping.

2.6.5 Land Use / Land Cover Mapping

The Land use/land cover map (Fig. 2.2) depicts the land use and land cover patterns on the surface of the earth on any desired scale. Land use refers to man's activities and the various uses which are carried on land. Land cover refers to natural vegetation, water bodies, rocks/soils, artificial cover and other resulting due to land transformations. The term land use and land cover are closely related and interchangeable. The purpose for which land is being used, commonly are associated with different types of cover such as forest, agriculture, wasteland, water bodies. This map is an important input in the derivation of other thematic maps from satellite remote sensing data.

Information on existing Land use / land cover and their spatial distribution forms the basis for any developmental planning. The current land use assessed for its suitability with the surface and ground water information in the light of land potential before suggesting alternate land use practices. The land use maps provide the up to date information on the type, location, spatial, distribution and extent of land use/ land cover.

Knowledge on the land use/land cover and its pattern of change promotes to understand present and the past land utilization pattern. Up to date information on the rate and kind of change is essential for proper land use planning and management of land resources for productive use, For instance the gross acreage estimates provided under different categories of land use/ land cover will be useful to the departments like agriculture, irrigation, revenue, forest & environment, state land use boards, town & country planning, National aforestation & eco-development board etc., for land use planning and decision making.

Methodology adopted for the development of data model on land use/ land cover on 1:10,000 scale is provided in this section. The data model is developed based on certain parameters namely, base line data, Cadastral maps, Central Survey Maps, SOI top sheet and Satellite data and ground truth data. Georeferencing of Satellite Data is very much necessary to prepare a LU/LC map and is performed as below:

- Acquisition of satellite data from NRSC in soft copy.
- Processing of satellite data has been done, so as to make data free from all the errors caused by atmosphere, geometry, and radiometry during the acquisition of data.
- Georeferencing of high resolution IRS P6 LISS IV-MX (5.8mt resolution) satellite data is done by using georeferenced toposheets and GCPs collected from field survey.
- After placing GCPs, input file and reference file are saved and after that, resampling of satellite data is carried. This procedure is repeated for every scene of concerned path and rows of satellite of the project.
- Overlaying of the georeferenced satellite data on respective toposheets has done to check whether the features are perfectly matched or not with toposheets, if some error results, repeat the above procedure of georeferencing.
- Georeferenced LISS- IV- MX (5.8 mt), and CARTOSAT-1 (2.5 mt) images are merged to get high resolution data (i.e. 2.6 Spatial resolution) with multispectral.
- Mosaic process of these scenes has done to make the image to show continuous terrain of project area.
- ArcGIS 9.1 software was used to fix 1:10,000 Scales. In the ArcGIS software without moving the scale, visible features in 1:10,000 scale has been digitized.

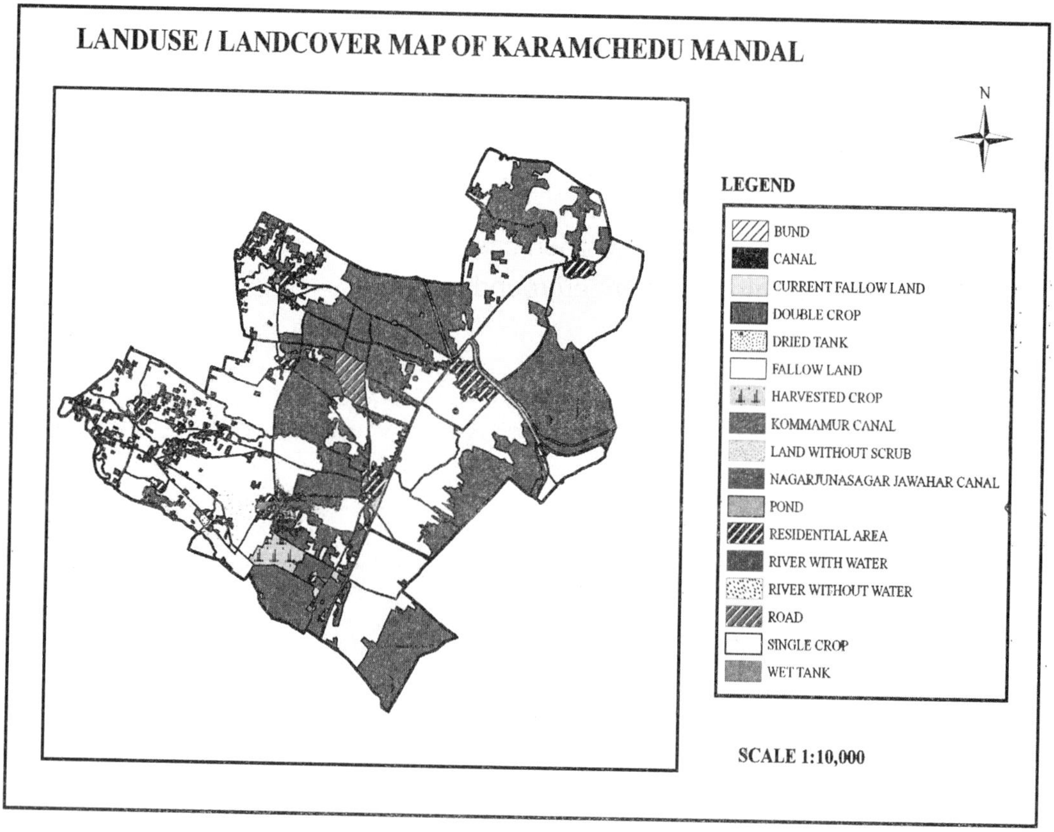

Fig. 2.2

REVIEW QUESTIONS

1. Write an essay on classification of Resources.
2. Explain Floods devastation due to improper land use practices.
3. Discuss the benefits and problems in Dams.
4. Discuss the environmental effects of mineral exploration.
5. Explain the remedial measures for the impacts on environment due to mineral resources.
6. Explain concept of land degradation.
7. Explain the issues related landslides.

8. Write short notes on
   - (i) Land use / Land cover mapping
   - (ii) Land as a resource
   - (iii) Droughts
9. Write short notes on
   - (i) Environmental impact on open cast mining
   - (ii) Ecological impacts of underground mining

Chapter 3

ENERGY RESOURCES

3.1 INTRODUCTION

Necessity for energy is ever-growing, as the transformation of energy is central to economic activity. Energy comes in many different forms, such as electrical, sound, light, chemical, heat (thermal), mechanical, and nuclear energy. The substantial rise in living standards experienced by developed countries over the last 250 years has depended to a significant extent on the exploitation of enormous amounts of energy. Present world energy consumption drawn from fossil fuels is approximately three quarters. These carbon rich materials produced over many millions of years are possible to be exhausted within a few hundred years, a few within decades. Energy is used by humans for transportation, industry, space heating, cooking, agriculture, and leisure. Most of the world's fiscal escalation has relied on non- renewable fossil fuel energy from oil, coal and natural gas resources since 1950.

Globally, the major users of fossil fuel energy are the more developed countries, which use this energy to run their industrialized economies. Table 5.2 depicts total and per capita consumption of electricity for the world and some selected countries. The data presented here suggests that energy use varies significantly between countries. The table indicates that the per capita consumption of electricity of the selected 'more developed countries (United States, Canada, and Japan) is some 10 to 75 times greater than less developed nations' (China, India, and Uganda). Similar differences also exist with the world's other nations.

Running out of fuel will be a serious problem but the environmental effects of energy consumption are potentially worse. Most forms of energy, and fossil fuels in particular, have a major impact on the environment. This includes local acute problems such as photochemical smog, more widespread issues like acid rain and global concerns, particularly climate resulting from greenhouse gas emissions.

Table 3.1 Total and per capita consumption of electricity for the world and selected countries.

Region	Population (in millions)	Total Yearly Electricity Consumption (millions of kilowatt-hours)	Per Capita Electricity Consumption (kilowatt-hours)
World	6, 081	12,342,000 (1998 estimate)	2,030
United States	276 (2000 estimate)	3,365,000 (1998 estimate)	12,190
Canada	31 (2000 estimate)	551,000 (1998 estimate)	17,770
Japan	127 (2000 estimate)	926,000 (1998 estimate)	7,290
China	1,262 (2000 estimate)	1,014,000 (1998 estimate)	800
India	1,014 (2000 estimate)	416,000 (1998 estimate)	410
Uganda	23 (2000 estimate)	622 (1998 estimate)	27

(Data Source: CIA)

3.2 ENERGY DEMANDS

As per the International Energy Outlook 2005, world marketed energy consumption is projected to increase on average by 2.0 percent per year over the 23-year forecast horizon from 2002 to 2025—slightly lower than the 2.2-percent average annual growth rate from 1970 to 2002. Worldwide, total energy use is projected to grow from 412 quadrillion British thermal units (Btu) in 2002 to 553 quadrillion Btu in 2015 and 645 quadrillion Btu in 2025 (Figure 3.1). Emerging economies account for much of the projected growth in marketed energy consumption over the next two decades, with energy use in the group more than doubling by 2025. Strong projected economic growth drives the demand for energy use in the region.

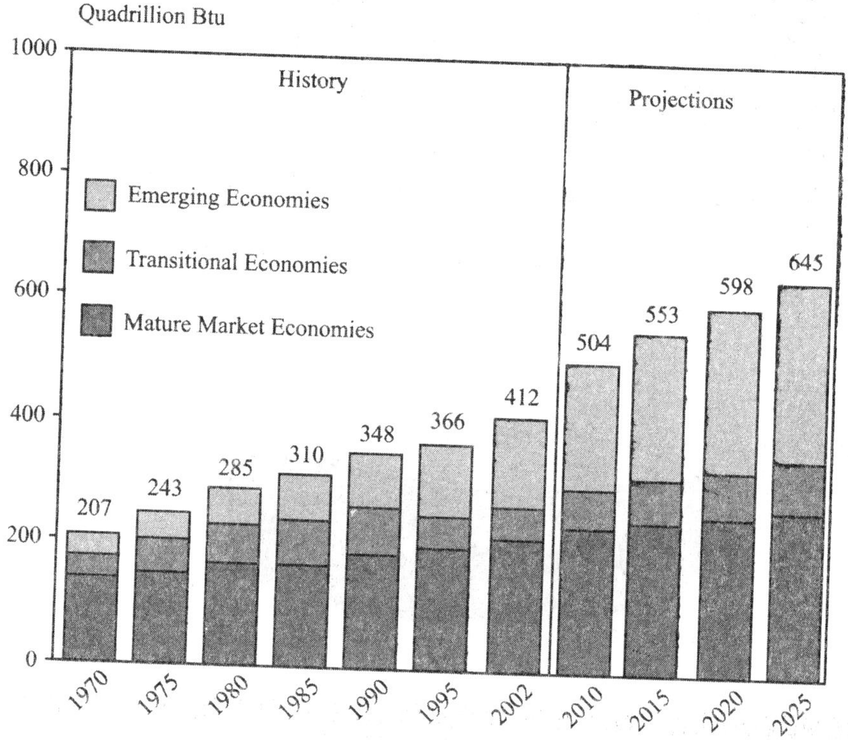

Fig. 3.1 World marketed energy consumption by Region, 1970-2025.

On a worldwide basis, the industrial and transportation sectors show the fastest growth in the *IEO2005* reference case, averaging 2.1 per cent per year in both sectors. Slower growth is projected in the residential and commercial sectors, averaging 1.5 per cent and 1.9 per cent per year, respectively, from 2002 to 2025 for the world as a whole. In the mature market economies, where population growth generally is slow or negative over the forecast, energy use in the commercial sector grows at a faster pace (1.3 per cent per year) than in any other end-use sector, based on expectations of rapid increases in the use of new telecommunications technologies and office equipment as these nations continue to shift to service economies. _World energy consumption was estimated to be 9670 Million Tonnes of Oil Equivalent (MTOE) in 1997, 80% of which was provided by fossil fuels. (Table 3.2).

Table 3.2 World energy sources by type - 1997:

Energy source	MTOE	% of world total
Fossil fuels	7721	79.8
Nuclear	624	6.5
Hydro	221	2.3
Combustible renewables & waste	1064	11.0
Other	40	0.4
World total	9670	100

Share of Total Primary Energy Supply in 2003

India

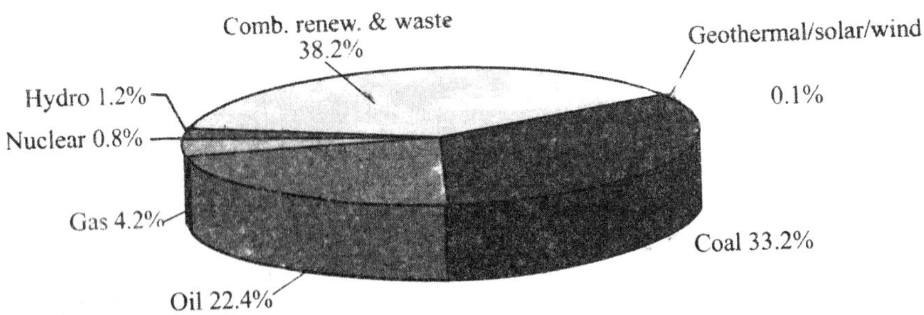

3.3 RENEWABLE AND NON-RENEWABLE ENERGY SOURCES

All forms of energy come from a "Source". For example, the sun is a source for light energy as well as solar energy, which can be changed into electricity through the use of solar panels. These sources can be sorted into two groups. The first is renewable (a type of energy that can be used over and over, like the sun). The second is non-renewable (an energy source that can be used up and is unable to renew - or create more of itself - in a short amount of time).

Renewable energy sources include the sun, wind, water, heat from within the Earth (called geothermal energy), and plant (biomass) waste products. These renewable sources are all natural-occurring elements. Human beings have found ways to use these energy sources. Besides harnessing the sun's solar energy, we use wind to power windmills, turbines to create hydroelectric energy with water, geothermal energy (heat from within the Earth) for use in power plants. Power plants are not strong bushes or weeds; a power plant is a large building where power is produced.

If an energy resource is being used faster than it can be replaced (for example, coal takes millions of years to form) then it will eventually run out. This is called a non-renewable energy source. Non renewable resources are finite in quantity and are being used faster than its ability to regenerate itself. Examples of non-renewable resources include metallic minerals, fossil fuels and nuclear power.

3.3.1 Renewable Energy Resources

The use of renewable resources is becoming more and more important in our society. It is intensively connected to natural bio-resources, agricultural production and new developments in the global agricurtural industry. In addition, environmental concerns especially about climate changes and about the use of fossil fuels and raw materials have increased the impact of renewable resources.

Renewable energy resources are Bio-energy(Biomass) - organic (plant, animal) materials, Geothermal - the heat of the earth, Hydroelectric - energy from flowing or falling water (streams, dams), Ocean energy - from waves, tides, or temperature differences in the ocean, Solar - from the sun (two types of systems: solar thermal and photovoltaic) and wind - from the wind. Renewable energy (sources) or RES capture their energy from existing flows of energy, from on-going natural processes, such as sunshine, wind, flowing water, biological processes, and geothermal heat flows.

Most renewable energy, other than geothermal and tidal power, ultimately come from the Sun. Some forms of storing solar energy such as rainfall and wind power, are considered short-term solar-energy storage, whereas the energy in biomass is accumulated over a period of months, as in straw, or through many years as in wood. Capturing renewable energy by plants, animals and humans does not permanently deplete the resource. Fossil fuels, while theoretically renewable on a very long time-scale, are exploited at rates that may deplete these resources in the near future.

Renewable energy resources may be used directly, or used to create other more convenient forms of energy. Examples of direct use are solar ovens, geothermal heating, and water and windmills. Examples of indirect use which require energy harvesting are electricity generation through wind turbines or photovoltaic cells, or production of fuels such as ethanol from biomass.

3.3.1.1 *Bioenergy*

Bioenergy uses renewable biomass resources to produce an array of energy related products including electricity, liquid, solid, and gaseous fuels, heat, chemicals, and other materials. Bioenergy accounts for about three per cent of the primary energy production. This come from biomass, i.e., any plant derived organic matter available on a renewable basis, including dedicated energy crops and trees, agricultural food and feed crops, agricultural crop wastes and residues, wood wastes and residues, aquatic plants, animal wastes, municipal wastes, and other waste materials.

Types of Bio-energy and Bio-fuels: Liquid fuels including ethanol, methanol, biodiesel, and gaseous fuels such as hydrogen and methane derived from biomass feedstocks. Biofuels are liquid fuels made from esters, alcohols, ethers, and other biomass chemicals. They are renewable fuels that can be produced in any climate using already developed agricultural practices. Common biofuels include: ethanol and biodiesel. Ethanol is made from starches or sugars, typically grain or corn. Biodiesel is an ester made from fats or oils. Cellulosic ethanol is the - future.

Advantages of Bio-fules:

- Because bio-fuels are renewable they can be used indefinitely without depleting earth's natural resource reserves
- Biofuels can be produced in short amounts of time (for example: one growing season) whereas non-renewables, like fossil fuels, take 40 million years or more to be produced.
- Biofuels are carbon-neutral, which means the net CO_2 outputs equal the net CO_2 inputs. Biofuels reduce harmful emissions into the atmosphere. It is renewable and does not contribute to global warming due to its closed carbon cycle. Carbon in the fuel was originally removed from the air by plants so there is no net increase in carbon dioxide levels. It provides substantial reductions in carbon monoxide, unburned hydrocarbons, and particulate emissions from diesel engines. Most emission tests have shown a slight oxides of nitrogen (NOx) increase with biodiesel. This increase in NOx can be eliminated with a small adjustment to the engine's injection timing while still retaining the particulate decrease. Biodiesel has excellent lubricating properties, when added to regular diesel fuel in an amount equal to 1-2%, it can convert fuel with poor lubricating properties, such as modern ultra-low-sulfur diesel fuel, into an acceptable fuel.
- Biodiesel be made from a variety of feedstocks:
    - Soyabean oil, corn oil, canola (an edible variety of gape seed) oil, cottonseed oil, mustard oil, palm oil, sunflower oil, linseed oil, Jatropha oil, etc.

- Restaurant waste oils such as frying oils
- Animal fats such as beef tallow or pork lard
- Trap grease (from restaurant grease traps), float grease (from waste water treatment plants), etc.

- Biofuels strengthen economy by:
- Reducing dependence on foreign oil (thereby reducing the trade deficit)
- Encouraging growth in the agricultural sector
- Biopower electricity generated from biomass. Based on direct-combustion technology: burning of biomass to produce steam in boilers. The steam is used to produce electricity in steam turbine generators. Most biopower produced is from waste wood. Future biopower technologies may include co-firing, gasification (biogas), pyrolysis, and anaerobic digestion.
- Biobased chemicals and industrial products, other than food and feed, derived from biomass feedstocks. Examples: green chemicals, renewable plastics, natural fibers, and natural structural materials.

3.3.1.2 *Geothermal Energy*

Developments in alternative energy sources sparked by threats of traditional energy resource exhaustion, drive to self-sufficiency,and drive to find alternative energy sources that are widely available, versatile, renewable, and have limited impact on the environment.

Geothermal Energy is the energy heat generated by natural processes occurring within the earth. Fumroles, hot springs and mud pots are natural phenomena that result from geothermal activity. Internal heat from the earth (which is produced by decay of natural radioactive materials). Most likely sites are near plate boundaries with active volcanoes and high heat flow., e.g. Pacific Rim, Iceland, Mediterranean. Facilities for geothermal energy exploitation are widely in use in Italy, US, Japan, NZ, Mexico, USSR.

Traditional utilization of geothermal energy: Natural releases of geothermal energy have been utilized for centuries in Balneology (Healing, Hygiene), Domestic services like Cooking, laundry, (Ex. Native New Zealanders), Mineral extraction, where geothermal water can contain useful minerals like boric acid, sulfur, vitriol or aluminum.

Harnessing of geothermal energy: Temperatures in the earth vary with depth as shown in Fig. 3.2. Within the earth, different areas have different thermal gradients and thus different utilization potentials. Higher thermal gradients correspond to areas containing more geo-thermal energy. Geothermal areas that can be used for large scale operations such as power generation require specific thermal gradients. Areas possessing these gradients are classified as geothermal fields and are located only in selected areas of the globe. Geothermal fields are the thermal areas, where permeable rock formations below ground contain a working fluid without which the area could not be exploited on a large scale.

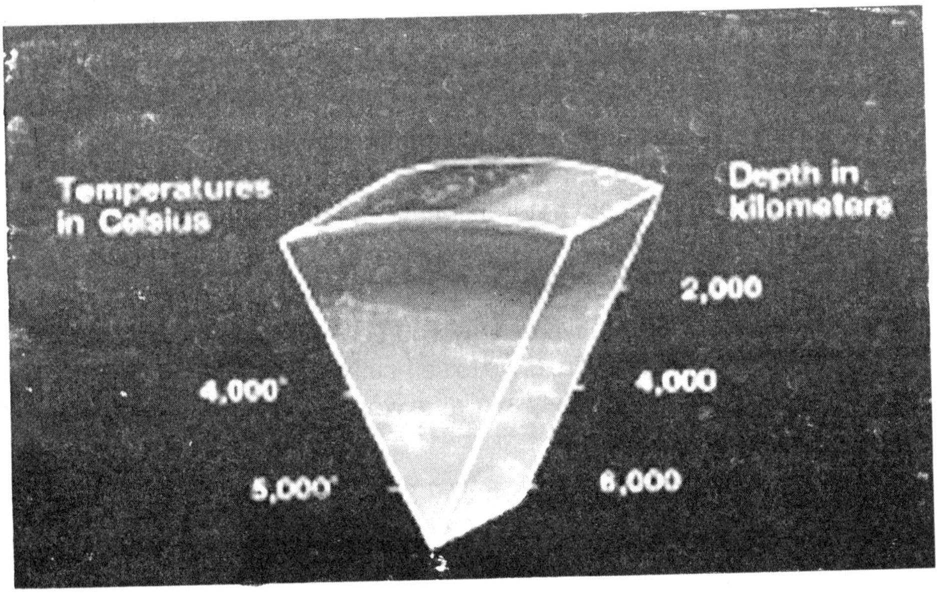

Fig. 3.2 Temperatures in the Earth.

Geothermal field characterizations:

- Semi-thermal field- produces water up to 100 °C from drilling depths of 1-2 km
- Wet hyper-thermal field (water-dominated)- produces pressurized water > 100 °C
- Dry hyper-thermal field (vapor-dominated)- produces dry saturated, or slightly superheated steam at $P > P_{atm}$

By exploiting geothermal fields, particularly hyper-thermal fields, geothermal energy can be harnessed on a large scale. Semi-thermal fields typically found in areas having abnormally high temperature gradients, Hyper-thermal fields generally located at tectonic plate boundaries in seismic zones. Heat flows outward from the center as a result of radioactive decay. The crust (about 30 and 60 km thick), insulates us from the interior heat, a solid inner core followed by liquid outer core, with the mantle by semi-molten state, and temperature at base of crust about 1000 °C, increases slowly into the core. The hot spots are located 2 to 3 km form the surface.

Tectonic plates, are in constant motion (several centimeters/yr). When collision or grinding occurs, it can create mountains, volcanoes, geysers and earthquakes. Near the junctions of these plates, is where geothermal heat travels rapidly from interior ? Distribution of major geothermal energy reserves is shown in Fig. 3.3.

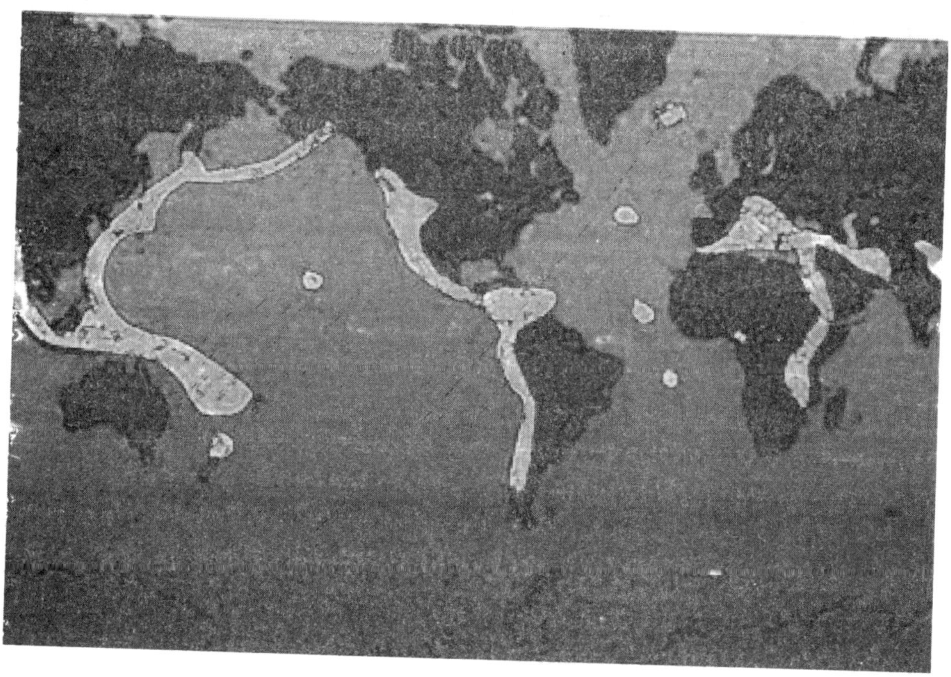

Fig. 3.3 Hottest known Geothermal Regions.

- Environmental impacts for the installation of geo-thermal power plants are much less than traditional power plants in regards to land impacts, air impacts, surface and groundwater impacts and aesthetic impacts further reduced in systems where geothermal waste water and steam is reinjected into the ground.

Severity of environmental impacts dependant on: type of thermal resource developed, chemical composition of geothermal fluid, chemical composition of subsurface rock, geology, hydrology and topography of the area along with the technology used for energy production and pollution control. Management planning can often reduce the affects of pollution through emission controls and proper planning.

SO_2 and CO_2 Emissions

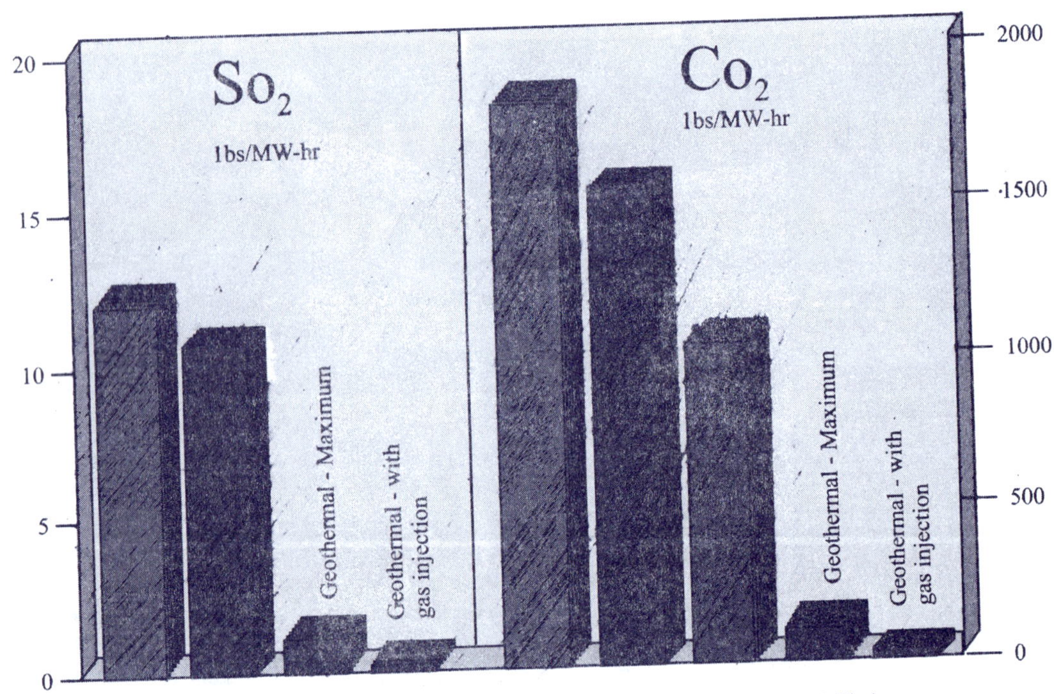

After Goddard & Goddard, GRC Transactions, 14,643-649 (1lb = 0.4536 kg)

3.3.1.3 *Hydroelectric Energy*

Hydropower must be one of the oldest methods of producing power. Hydropower energy is obtained from flowing water. Energy in water can be harnessed and used, in the form of motive energy or temperature differences. The most common application is the dam, but it can be used directly as a mechanical force or a thermal source/sink.

Hydroelectric power from potential energy of the elevation of waters, now supplies about 715,000 MWe or 19% of world electricity, and large dams are still being designed. Apart from a few countries with an abundance of it, hydro power is normally applied to peak-load demand, because it is so readily stopped and started. Nevertheless, hydroelectric power is probably not a major option for the future of energy production in the developed nations because most major sites within these nations with the potential for harnessing gravity in this way are either already being exploited or are unavailable for other reasons such as environmental considerations.

Small scale hydro or micro-hydro power has been increasingly used as an alternative energy source, especially in remote areas other power sources are not viable. Small scale hydro power systems can be installed in small rivers or streams with little or no discernable environmental effect on things such as fish migration. Most small scale hydro power systems make no use of a dam or major water diversion, but rather use water wheels with little environmental impact. Water is needed to run a hydroelectric generating unit. It's held in a reservoir or lake behind the dam, and the force of the water being released from the reservoir through the dam spins the blades of a turbine. The turbine is connected to the generator that produces electricity. After passing through the turbine, the water reenters the river on the downstream side of the dam. (Fig. 3.4).

3.3.1.4 *Active Solar Heating System*

Active Solar Heating Systems-heated fluid is artificially circulated. Flat plate collector-flat metal plate absorbs sun's energy. Fluid comes in contact with plate and is circulated to needed place. Plate is contained in insulated box with glass lid (glass is opaqued to infrared re-radiation but allows in 90% of incident radiation).

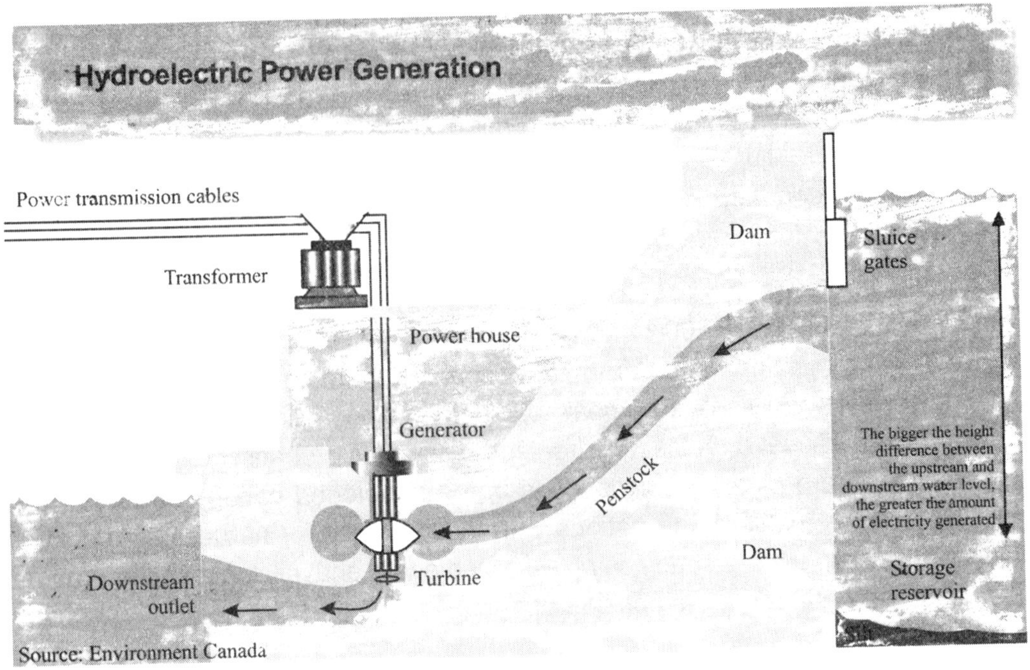

Photo courtesy U.S. Bureau of Reclamation
Outflow during a test at the hydropower plant at the Hoover Dam, located on the Nevada-Arizona border

Fig. 3.4 Hydroelectric power generation.

Types of Collectors

1. Tubes sandwiched between plate
2. Water trickle over plate
3. Black rubber mat with tubes and fins (low temp swimming pools)
4. Collector efficiency = 100% x (useful energy delivered)/(insolation on collector) number can be as high as 60-70%

Factors affecting efficiency

1. Water temp - since conduction losses depend on T, bigger T = more losses
2. Radiation losses - hot things radiate. Coating absorbers helps copper oxide film—absorbtance =.9, emissivity = .15
3. Angle of collector - depends on usage.

Storage: Several different types of storage systems exists, usage depends on space.

Volume heat capacity = amount of energy needed to raise one volume unit of material, one degree temperature = specific heat x density Ex. iron 1/8 heat capacity of water, but 8 times denser

We can use water under rock layers, especially for air system Phase-changing materials - heat of fusion release. can be smaller storage, but holds storage at specific temperature. Ex. Eutectic salts.

Uses

1. Space heating - Baseboard radiators. Heat from collector is pumped to storage tank. Fluid is then pumped out, and if needed, additional heat is added before going to baseboards
2. Hot water - Same as space heating, except water is definitely used (heat exchanger in storage tank).

Focused Collectors: Focused Collectors - an active solar system that uses curved mirrors to focus sunlight on working fluid. Can achieve temperatures greater than 180 F and up to 1000 F. Major use is in steam generators (why would you need 1000 F water or air?)

Passive Solar Heating Systems: Passive solar heating system - heated fluid is not artificially transmitted. Natural means (convection and conduction) are used to do all of the transport that is necessary. Big gain in savings. This type of system uses the fact that the amount of solar energy transmitted through glass over 24 period is greater than heat lost through them. All types need excellent insulation, solar collection and thermal storage facilities. Four common types are

- Direct gain - direct sunlight heats room. Needs thermal mass to store heat (Concrete, rocks, etc). Adobe houses of southwest
- Indirect gain - collect and store energy in one part and allow natural convection to transfer energy to other parts. Ex. Trombe wall
- Attached greenhouse - much like indirect gain. However also provides barrier during summer of direct sunlight in living quarters. Also good for food production
- Thermosiphon - can be used for hot water. For home heating or window unit uses natural bouyancy to heat

Economics: Active systems are expensive, passive less so. More expensive to retro-fit than to build in. In these times, with no incentive to do it (energy prices low, no solar dividend) and economy way it is, nobody thinks about it. Biggest push may be more due to environmental reasons.

- Possible savings — 25% of energy usage goes to heating and cooling
- Northern states have a greater need of hot air in winter, but receives less insolation than southern states
- South's biggest use probably for hot water. Domestic hot water accounts for 4% of energy usage.
- Batteries store the power generated and discharge the power as needed.
- Battery bank consists of one or more solar deep-cycle type batteries.
- Depending on the current and voltages for certain applications the batteries are wired in series and/or parallel.

Three ways to convert sunlight into electricity, mainly, photovoltaics, wind turbines and solar thermal (steam) turbines .

Solar Cell Principles: Photoelectric effect - discovered by Heinrich Hertz in 1887. Explained by Einstein in 1905. Electrons are emitted when light strikes metals. Puzzle was that for certain colors of light, no electrons emitted. Explanation - Light has wave and particle characteristics. If we think of particle, then each photon has energy of $E=hf$. As photon is absorbed by metal, if hf is greater than binding energy of electrons to metal, then electrons are liberated.

Solar Cell Manufacture **:** Most solar (PV) cells consist of two semiconductor materials joined together. Silicon is "doped" with phosphorous to create an n-type semiconductor crystal, which is joined to silicon "doped" with boron (p-type semiconductor crystal) to create a p-n junction. This creates a potential barrier that "gives direction" to the liberated electrons, i.e. the liberated electrons are driven in the direction of the potential energy drop.

p-n junctions can also be formed out of amorphous silicon (no crystalline structure). Dangling bonds (lack of crystalline structure) can capture free electrons. These are cheap to manufacture and are efficient under fluorescent light. Materials other than silicon can be used to create the p-n junctions. Materials like gallium arsenide, cadmium telluride, and cadmium sulfide can be used. Efficiencies higher than silicon-based PV cells can be achieved (books quote of 40% is out of line with long term usage; best efficiencies are about 20-25%).

3.3.1.5 *Wind Energy*

Wind power is the kinetic energy of wind, or the extraction of this energy by wind turbines. In 2004, wind power became the least expensive form of new power generation, dipping below the cost per kilowatt-hour of coal-fired plants. Wind power is growing faster than any other form of electrical generation, at about 37%, up from 25% growth in 2002. In the late-1990s, the cost of wind power was about five times what it is in 2005, and that downward trend is expected to continue as larger multi-megawatt turbines are mass-produced.

An estimated 1 to 3 percent of the energy from the Sun is converted into wind energy. This is about 50 to 100 times more energy than what is converted into biomass by all the plants on earth through photosynthesis. Most of this wind energy can be found at high altitudes where continuous wind speeds of over 160 km/h (100 mph) are common. Eventually, the wind energy is converted through friction into diffuse heat all through the earth's surface and atmosphere.

While the exact kinetics of wind are extremely complicated and relatively little understood, the basics of its origins are relatively simple. The earth is not heated evenly by the sun. Not only do the poles receive less energy from the sun than the equator does, but dry land heats up (and cools down) more quickly than the seas do. The differential heating powers a global atmospheric convection system reaching from the earth's surface to the stratosphere acts as a virtual ceiling. The change of seasons, change of day and night, the Coriolis effect, the irregular albedo (reflectivity) of land and water, humidity, and the friction of wind over different terrain are some of the many factors which complicate the flow of wind over the surface.

3.3.2 NON-RENEWABLE ENERGY RESOURCES

Most of the world's **energy** comes from the burning of oil. Humans use **oil** for the purposes like transportation, industry, agriculture, residential uses and electric utilities.

3.3.2.1 ***Crude oil*** (or petroleum) is composed of more than a hundred different **hydrocarbon** compounds. The refining process separates crude oil into many different commercial products including gases, gasoline, heating oil, diesel oil, and asphalt. Crude oil also contains petrochemicals which are compounds that are used in the production of **fertilizers**, plastics, paints, **pesticides**, medicines, and synthetic fibers like nylon.

What are fossil fuels?

- Derived from the remains of once living organisms.
- Formed some 500-200 million years ago.
- The three major fossil fuels are oil, coal, and natural gas.
- Currently being consumed at a rate faster than they are being produced.

3.3.2.2 ***Fossil fuels*** are hydrocarbons that are pumped out of the ground. Fossil fuels encompass **coal, oil** and **natural gas**. They are formed over a millennia of time so producing more is not possible. Carbon cycle is shown in the following picture. (Fig. 3.5).

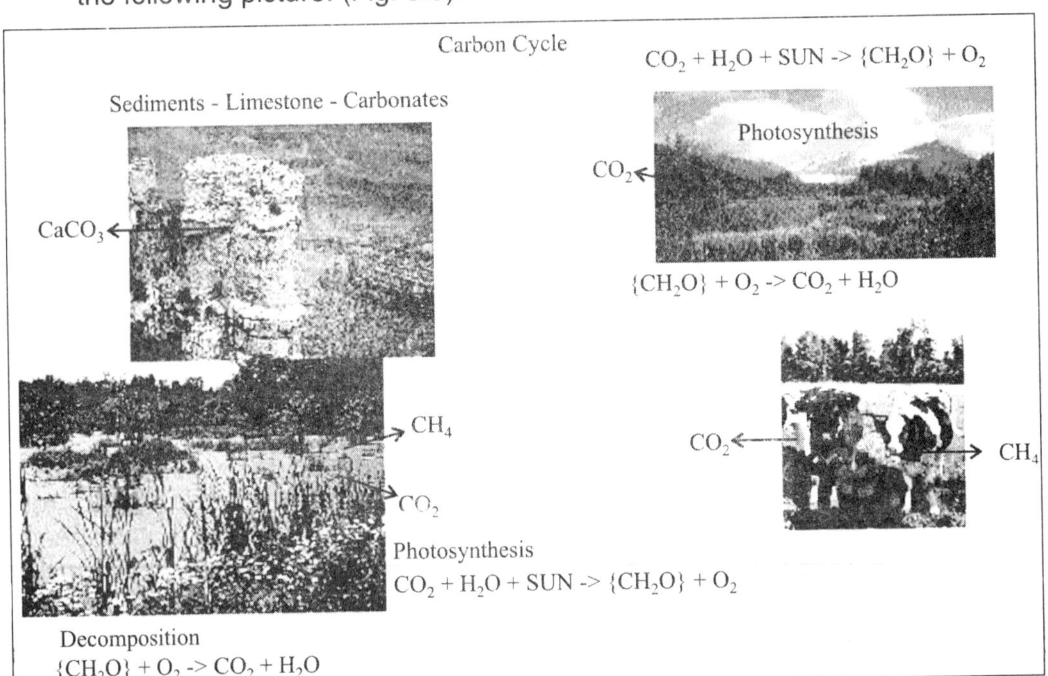

Fig. 3.5 Carbon cycle.

Examples are: Butane = C4H10, Hexane = C6H14, Octane = C8H18.

Molecular weight has much to do with function & properties

1 carbon unit	methane gas
3 carbon units	propane gas used for outdoor grills
4 carbon units	butane liquid – used i n cigarette lighters
8 carbon units	octane – gasoline
15-17 carbon units	mineral oil
22-30 carbon units	paraffin wax - candle
10,000 carbon units	usable, moldable plastic like a milk jug

World reserves of fossil fuels (as of 2000)

Coal 250 year supply

Petroleum 45 year supply

Natural gas 70 year supply

3.3.2.3 *Natural Gas*

Most efficient and cleanest burning, leaving no residues. It has the highest energy output for its mass. Often found with gas, but also can be found by itself in the ground. Typically contains 60-80% methane with some propane and butane. Natural gas is purified before using to remove CO_2, containing compounds & water vapor.

In its underground gaseous state, natural gas is a mixture of 50 to 90% by volume of methane gas (CH_4) and smaller amounts of heavier gaseous hydrocarbon compounds such as propane (C_3H_8) and butane (C_4H_{10}). Propane and butane are separated from natural gas and are stored as liquefied petroleum in special pressurized tanks. These two substances are primarily used as fuel for heating and cooking. Methane is more important economically. Methane is burned in furnaces to supply residential and commercial buildings with heat, used to heat water in hot water tanks, used in power plants to generate electricity, and used in a variety of ways in the organic chemistry industry.

Several non-hydrocarbon gases are also found in natural gas mixtures. Nitrogen dioxide, carbon dioxide, and hydrogen sulfide may be found in substantial proportions. Nitrogen is inert and non-combustible. If present in significant quantities, it reduces the heating value of the natural gas mixture and must be separated before the fuel is used for the commercial purposes.

Carbon dioxide is another unwanted component of the gas mixture. Normally, it is removed from the mixture to raise heating value, reduce volume, and support even combustion properties. Hydrogen sulfide is removed from natural gas by a process similar to that used in crude oil refining.

Two types of natural gas are commonly extracted from the lithosphere. Conventional natural gas is found in deposits that occur above oil. If the natural gas occurs in bedrock by itself it is called unconventional natural gas. Conventional supplies of natural gas are better distributed and believed to be larger than those of crude oil. World supply of known reserves and undiscovered, economically recoverable occurrences of conventional natural gas are estimated at 50 years at present consumption rates.

Coal: Fuel is any type of material that stores energy and can be burned to supply heat. Fossil fuels are types of organic fuel that formed from the remains of plants or animals (like dinosaurs) that are buried in rock layers all over the Earth. Heat from inside the Earth, along with pressure from rock and soil on the remains (or "fossils"), caused them over time to become coal, petroleum, or natural gas. Coal is a solid in the form of a soft, easy-to-burn rock; petroleum is a liquid, and natural gas is more like a vapor, not gasoline that you use as fuel for cars and trucks. Fossil fuels contain chemical energy that is released by burning.

Coal is formed in several stages as the remains of plants are subjected to intense heat and pressure over millions of years. It is a complex mixture of organic compounds, with 30 % to 98 % carbon by weight plus varying amounts of water and small amounts of nitrogen and sulfur. Physical properties of varieties of coal is given in Table 3.3. Vegetation decomposes to peat. When peat becomes deeply buried with high pressure it becomes lignite. Over thousands of years, higher pressures and temperatures turns lignite into bituminous coal (soft coal). In mountainous areas, higher pressures make anthracite (hard coal, 90% carbon). (Table 3.4).

Table 3.3 Physical properties of varieties of coal.

Property	Peat	Lignite	Bituminous	Anthracite
Appearance	Brown	Brown	Brown black	Black
Hardness	Porous	Soft	Soft	Hard
Water Content %	~70	30-40	10	5
Ash	-	5	4	5
Sulfur	-	0.6	0.3	0.8
Carbon	50	41	50	85-95

Table 3.4 Coal Formation.

	Peat -------> Lignite -------> Subbituminous -------> Bituminous ------> Anthracite
Energy Content	Low --> High
Moisture Content	High ---> Low
Sulfur Content	High --> Low

Anthracite: Anthracite is coal with the highest carbon content, between 86 and 98 percent, and a heat value of nearly 15,000 BTUs-per-pound. Most frequently associated with home heating, anthracite is a very small segment of the U.S. coal market. There are 7.3 billion tons of anthracite reserves in the United States.

Bituminous: The most plentiful form of coal in the United States, bituminous coal is used primarily to generate electricity and make coke for the steel industry. The fastest growing market for coal, though still a small one, is supplying heat for industrial processes. Bituminous coal has a carbon content ranging from 45 to 86 percent carbon and a heat value of 10,500 to 15,500 BTUs-per-pound.

Subbituminous: Ranking below bituminous is subbituminous coal with 35-45 percent carbon content and a heat value between 8,300 and 13,000 BTUs-per-pound. Reserves are located mainly in a half-dozen Western states and Alaska. Although its heat value is lower, this coal generally has a lower sulfur content than other types, which makes it attractive for use because it is cleaner burning.

Lignite: Lignite is a geologically young coal which has the lowest carbon content, 25-35 percent, and a heat value ranging between 4,000 and 8,300 BTUs-per-pound. Sometimes called brown coal, it is mainly used for electric power generation.

About 60 % of the coal extracted in the world is burned in boilers to produce steam to generate electrical power. The rest is converted to coke, used to make steel, burned in boilers to produce steam, and used in various manufacturing processes. Electricity is produced at a coal-fired fossil plant by the process of heating water in a boiler to produce steam. The steam, under tremendous pressure, flows into a turbine, which spins a generator to produce electricity. Each boiler heats the water to about 1,000 degrees Fahrenheit (540 degrees Celsius) to create steam. The steam is sent into the turbines at pressures of more than 1,800 pounds per square inch (130 kilograms per square centimeter). This one plant generates about 10 billion kilowatt-hours a year, or enough electricity to supply 700,000 homes.

3.4 NUCLEAR ENERGY

Nuclear power is the power that is released by manipulating atoms, by either splitting them apart (fission) or fusing them together. (fusion) (Fig. 3.6). Finally, there are some minerals included as energy sources in the non-renewable category. A mineral such as uranium can be used to create heat and even electricity by means of nuclear fission (the splitting of its atoms).

Nuclear power accounts for about 20 per cent of the total electricity generated in the United States. Nuclear power plant operates basically the same way as a fossil fuel plant, with one difference: the source of heat. The process that produces the heat in a nuclear plant is the fissioning or splitting of uranium atoms. That heat boils water to make the steam that turns the turbine-generator, just as in a fossil fuel plant. The part of the plant where the heat is produced is called the reactor core.

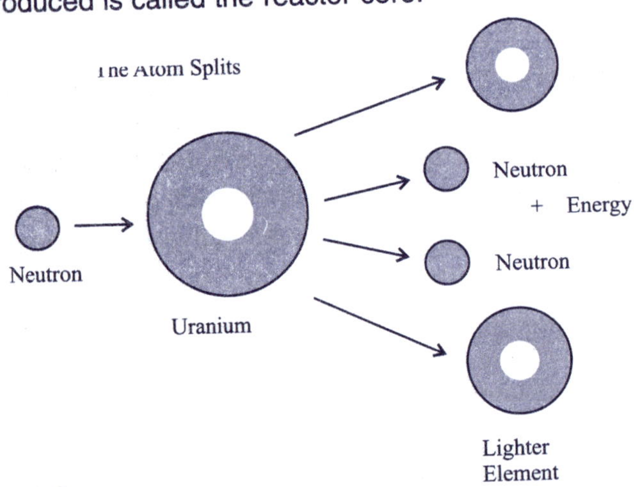

Fig. 3.6 Carbon cycle.

Most power plants burn fuel to produce electricity, but not nuclear power plants. Instead, nuclear plants use the heat given off during fission as fuel. Fission takes place inside the reactor of a nuclear power plant. At the center of the reactor is the core, which contains the uranium fuel.

The uranium fuel is formed into ceramic pellets. The pellets are about the size of your fingertip, but each one produces the same amount of energy as 150 gallons of oil. These energy-rich pellets are stacked end-to-end in 12-foot metal fuel rods. A bundle of fuel rods is called a fuel assembly.

Fission generates heat in a reactor just as coal generates heat in a boiler. The heat is used to boil water into steam. The steam turns huge turbine blades. As they turn, they drive generators that make electricity. Afterwards, the steam is changed back into water and cooled in a separate structure at the power plant called a cooling tower. The water can be used again and again.

In the BWR, the water heated by the reactor core turns directly into steam in the reactor vessel and is then used to power the turbine-generator. In a PWR, the water passing through the reactor core is kept under pressure so that it does not turn to steam at all — it remains liquid. Steam to drive the turbine is generated in a separate piece of equipment called a steam generator. A steam generator is a giant cylinder with thousands of tubes in it through which the hot radioactive water can flow. Outside the tubes in the steam generator, non-radioactive water (or clean water) boils and eventually turns to steam. The clean water may come from one of several sources: oceans, lakes or rivers. The radioactive water flows back to the reactor core, where it is reheated, only to flow back to the steam generator. Roughly seventy percent of the reactors operating in the U.S. are PWR.

Nuclear reactors are basically machines that contain and control chain reactions, while releasing heat at a controlled rate. In electric power plants, the reactors supply the heat to turn water into steam, which drives the turbine-generators. The electricity travels through high voltage transmission lines and low voltage distribution lines to homes, schools, hospitals, factories, office buildings, rail systems and other users.

REVIEW QUESTIONS

1. Write a short notes on Energy Demand.
2. Explain renewable and non-renewable sources of energy.
3. Explain Bioenergy, Geothermal energy.
4. What is meant by active solar heating system?
5. Write notes on (i) Solar cell principles, (ii) Wind energy and (iii) Natural gas

CHAPTER 4

BIODIVERSITY AND CONSERVATION

4.1 INTRODUCTION

The word 'biodiversity' is a combination of two words: biological and diversity. It refers to the variety of life on Earth and encompasses all the living things that exist in a certain area, in the air, on land or in water: plants, animals, microorganisms, fungi. The area considered may be as small as your backyard compost heap or as big as our whole planet. Animals and plants don't exist in isolation. All living things are connected to other living things and to their non-living environment (earth forms, rocks and rivers). If one tiny species in an ecosystem becomes extinct, we may not notice, or think it's important. But the biodiversity of that ecosystem will be altered, and all the ecosystems that the species belonged to will be affected.

Biological diversity therefore refers to the variety and variability among living organisms and the ecological complexes in which they occur. Diversity can be defined as the number of different items and their relative frequency. For biological diversity, these items are organized at many levels, ranging from complete ecosystems to the chemical structures that are the molecular basis of heredity. Thus, the term encompasses different ecosystems, species, genes, and their relative abundance.

Biological diversity is defined as the variability among living organisms from all sources including terrestrial, marine and other aquatic ecosystems and the ecological complexes of which they are part; this includes diversity within species, between species and of ecosystems. There is crop biodiversity, agri–horticultural biodiversity, herbal biodiversity, forest biodiversity all occurring at molecular, micro, and macro levels, a synergistic understanding alone will make us realize about the economic value of biodiversity resources.

4.2 TYPES OF BIODIVERSITY

There are three aspects to biodiversity: species diversity, genetic diversity and ecosystem/habitat diversity. All three interact and change over time and from place to place.

4.2.1 Habitat Diversity

Life exists in soil, air, water and other such habitats and biotic communities are controlled by the environmental variables. Ecosystems include not only the plants, animals and microbes that co-exists at a place but also the ways in which they interact with each other and adopt to their physical environment. For example the river Ganges includes the fish, aquatic insects, mussels and variety of plants that have adapted.

Habitat Diversity is the aggregate of different environment types in a region. For example, a country on the coast could have a diversity of ecosystems from sandy beaches with salt water adapted biodiversity, to lush rain forests and dry deserts, each with a unique set of species. Living organisms and physical environment interact with each other in an ecosystem.

4.2.2 Species Diversity

Biodiversity is the sum of the variety of all living organisms at the species level. It includes earth's plants, animals, fungi and microbes. Around 1.5 millions species of living organisms are named. There is a hidden wealth of 10 to 100 million more living organisms which have to be estimated. Species diversity is a function of both species richness and evenness. It measures the number of species in a given community and also distribution of each species within the community.

4.2.3 Genetic Diversity

Tremendous amount of genetic diversity exists within individual species. This genetic variability is responsible for the different characters in species. Such species adopt to different environments. Genetic diversity is the raw material from which new species arise through evolution. Genetic diversity refers to the variation at the level of individual genes besides providing a mechanism.

For ecological adaptation, more the variation, better the chances that at east some of the individuals will have an allelic variant that is suited for new environment. Such process will produce offspring with the variant that will in turn reproduce and continue the population into subsequent generations. Genetic diversity can be assessed at three levels:

(a) Diversity with in breeding populations.

(b) Diversity between breeding population.

(c) Diversity within species.

Genetic diversity also refers to the variation of genes within species or variation in the genetic composition of individuals within or among species, varieties or breeds.

4.3 VALUE OF BIODIVERSITY

The value of biodiversity in terms of its commercial utility, ecological services, social and aesthetic value is enormous. We get benefits from other organisms in innumerable ways. Sometimes we realize and appreciate the value of the organism only after it is lost from this earth. The multiple uses of biodiversity or biodiversity value may be classified as follows:

4.3.1 Consumptive use Value

These are direct use values where the biodiversity products can be harvested and consumed directly e.g. fuel, food, drugs, fibre etc.

(a) ***Food:*** A large number of wild plants are consumed by human beings as food. About 80,000 edible plant species have been reported from the wild. About 90% of present-day food crops have been domesticated from wild tropical plants.

(b) ***Drugs and medicines:*** About 75% of the world's population depends upon plants or plant extracts for medicines. The wonder drug Penicillin used as an antibiotic is derived from a fungus called *Penicillium.* Likewise, we get Tetracycline from a bacterium. Quinine, the cure for malaria is obtained from the bark of Cinchona tree, while Digitalin is obtained from foxglove (Digitalis) which is an effective cure for heart ailments. Recently Vinblastin and Vincristine, two anticancer drugs, have been obtained from Periwinkle *(Catharanthus)* plant, which possesses anticancer alkaloids. A large number of marine animals are supposed to possess anti-cancer properties which are yet to be explored systematically.

(c) ***Fuel:*** Our forests have been used since ages for fuel wood. The fossil fuels like coal, petroleum and natural gas are also products of fossilized biodiversity. Firewood collected by individuals are not normally marketed, but are directly consumed by tribes and local villagers, hence falls under consumptive value.

4.3.2 Productive use Values

These are the commercially usable values where the product is marketed and sold. It may include number of wild gene resources that can be traded for use by scientists for introducing desirable traits in the crops and domesticated animals. These may include the animal products like tusks of elephants, musk from musk deer, silk from silkworm, wool from sheep, fir of many animals, lac from lac insects etc, all of which are traded in the market. Many industries are dependent upon the productive use values of biodiversity e.g. the paper industry, plywood industry, railway sleeper industry, silk industry, textile industry, ivory-works, leather industry, pearl industry etc.

4.3.3 Social Value

These are the values associated with the social life, customs, religion and psycho-spiritual aspects of the people. Many of the plants are considered holy and sacred in our country like Tulsi (holy basil), Peepal, Mango, Lotus, Bael etc. The leaves, fruits or flowers of these plants are used in worship. Thus social life, songs, dances and customs are closely woven around the wildlife. Many animals like cow, snake, bull, peacock, owl etc. also have significant place in our psycho-spiritual arena and thus hold special importance. Thus, biodiversity has distinct social value, attached with different societies.

4.3.4 Ethical Value

It is also sometimes known as existence value. It involves ethical issues like *"all life must be preserved"*. It is based on the concept of *"Live and Let Live"*. If we want our human race to survive, then we must protect all biodiversity, because biodiversity is valuable. The ethical value means that we may or may not use a species, but knowing the very fact that this species exists in nature gives us pleasure. We all feel sorry when we learn that "passenger pigeon' or "dodo" is no more on this earth. We are not deriving anything direct from kangaroo, zebra or giraffe, but we all strongly feel that these species should exist in nature. This means, there is an ethical value or existence value attached to each species.

4.3.5 Aesthetic Value

Great aesthetic value is attached to biodiversity. No one of us would like to visit vast stretches of barren lands with no signs of visible life. People from far and wide spend a lot of time and money to visit wilderness area where they can enjoy the aesthetic value of biodiversity and this type of tourism is now known as *eco-tourism.* The "willingness to pay" concept of such eco-tourism gives us even a monetary benefit for aesthetic value of biodiversity. Eco-tourism is estimated to generate 12 billion dollars of revenue annually that roughly gives the aesthetic value of biodiversity.

4.3.6 Option Values

These values include the potentials of biodiversity that are presently unknown and need to be explored. There is a possibility that we may have some potential cure of AIDS or cancer existing within the depths of marine ecosystem, or a tropical rainforests.

Thus option value is the value of knowing that there are biological resources existing on this biosphere that may one day prove to be an effective option for something important in the future. Thus, the option value of biodiversity suggests that any species may prove to be a miracle species some day. The biodiversity is like precious gifts of nature presented to us. We should not commit the folly of losing these gifts even before unwrapping them. The option value also includes the values, in term of the option to visit areas where a variety of flora and fauna, or specifically some endemic, rare or endangered species exist.

4.3.7 Ecosystem Service Value

Recently, a non-consumptive use value related to self maintenance of the ecosystem and various important ecosystem services has been recognized. It refers to the services provided by ecosystems like prevention of soil erosion, prevention of floods, maintenance of soil fertility, cycling of nutrients, fixation of nitrogen, cycling of water, their role as carbon sinks, pollutant absorption and reduction of the threat of global warming etc.

4.4 THREATS TO BIODIVERSITY

Extinction is a natural event and, from a geological perspective, routine. We now know that most species that have ever lived have gone extinct. The average rate over the past 200 my is 1-2 species per year, and 3-4 families per my. The average duration of a species is 2-10 million years (based on last 200 million years). There have also been occasional episodes of mass extinction, when many taxa representing a wide array of life forms have gone extinct in the same blink of geological time. In the modern era, due to human actions, species and ecosystems are threatened with destruction to an extent rarely seen in earth history. Probably only during the handful of mass extinction events have so many species been threatened, in so short a time.

First, we can attribute the loss of species and ecosystems to the accelerating transformation of the earth by a growing human population. As the human population passes the six billion mark, we have transformed, degraded or destroyed roughly half of the word's forests. We appropriate roughly half of the world's net primary productivity for human use. We appropriate most available fresh water, and we harvest virtually all of the available productivity of the oceans. It is little wonder that species are disappearing and ecosystems are being destroyed.

Second, we can examine six specific types of human actions that threaten species and ecosystems - the "sinister sextet" Viz., over-hunting, habitat destruction, invasion of non-native species, domino effects, pollution, and climate change.

Habitat loss presents the single greatest threat to world biodiversity, and the magnitude of this threat can be approximated from species-area curves and rates of habitat loss. The spread of non-native species threatens many local species with extinction, and pushes the world's biota toward a more homogeneous and widely distributed sub-set of survivors. Climate change threatens to force species and ecosystems to migrate toward higher latitudes, with no guarantee of suitable habitat or access routes. These three factors thus are of special concern.

4.5 CONSERVATION PROCEDURE

We must make every effort to preserve, conserve, and manage biodiversity. Protected areas, from large wilderness reserves to small sites for particular species, and reserves for controlled uses, will all be part of this process. Such systems of protected areas must be managed to take account of a range of ecological and human-induced changes. This is no small task; yet humans must be equal to this challenge, or risk becoming irrelevant.—Peter Bridgewater, National Parks and Wildlife Service, Australia

The vanishing biodiversity is to be carefully conserved for sustainability of the same. Following are general methods of conservation,

- In situ conservation
- Ex situ conservation
- Micro-Propagation or tissue culture
- Cultivation in National Parks, Sanctuaries
- Wild life projects ex., Tiger, lion, elephant projects
- Conservation of sacred species in Sacred groves.

Major strategies of biodiversity conservation after J S Singh 2002 are:

Strategy	Work element
Insitu	Establish protected area network, with appropriate management practices, corridors to link fragments restore degraded habitats within and out side Pas
Ex situ	Establish botanical and zoological gardens, conservation stands; banks of germ plasm, pollen, seed, seedlings, tissue culture, gene and DNA, etc..,
Reduction of biotic Pressure	Reduce anthropogenic pressure on natural population by cultivating them elsewhere.
Rehabilitation	Identify and rehabilitate threatened species; launch augmentation, reintroduction or introduction programmes.

known as Jhum in the Indian subcontinent is more common in the hilly regions. Due to the increase in population, people are compelled to cultivate on the same plot of land more frequently as there is very little forest area available. Forests are also being converted to permanent settlements. Thus, forests cannot regenerate, and, in some cases, forest areas have become wasteland within a few years due to frequent cultivation. For example, after independence India lost forest area in the following manner 4696 million hectares forest land to non-forestry purposes, 0.07 million hectares to illegal encroachment, 4.37 million hectares to cultivation, 0.518 million hectares to river valley projects, 0.141 million hectares to industries and townships, and 0.061 million hectares to transmission lines and roads.

To conserve forest areas, the government launched Joint forest management and Social forestry schemes, with some success. But other conservation measures have to be taken to save the forests from further depletion and degeneration. Governments and communities should take steps to plant trees on a large scale. The community should be consulted and trees that will meet their requirements should be planted. People living in the rural and forestry areas should be sensitized to the damage done to their surroundings by the felling of trees. They should be encouraged to cut branches, twigs and leaves of the trees for their daily requirements. Horticulture as an alternative source of income should be encouraged. All this, collectively, would halt the depletion of forest cover. In order to conserve the forest cover, the concerned governments have intiated number of socially relevant, environmentally friendly activities to develop a variety of forests like, Agrforestry, Social forestry, Farm forestry, Community forestry, Extension forestry etc. Farmers in India grow agricultural crops, rear animals, plants and certain trees on their land, often on the boundary area.

Agroforestry is defined by some as a dynamic, ecologically-based natural farm management system that, along with agriculture and the integration of trees on farms, has many environmental benefits. Agroforestry reduces the farmers' dependency on forests even as it provides them economic benefits. It results in more diverse, healthy, and sustainable land-use systems. It focuses on meeting the economic, environmental, and domestic needs of people on their private lands. For hundreds of years, farmers have nurtured trees in their fields, pasturelands, and around their homes. Social forestry also aims at raising plantations by the common man so as to meet the growing demand for timber, fuel wood, fodder etc., thereby reducing the pressure on the traditional forest area. This concept of village forests to meet the needs of the rural people is not new. It has existed through the centuries all over the country but it is now given a new character.

Farm Forestry are of two types, may be considered under Social forestry programmes. The two types are both commercial and non-commercial farm forestry. In commercial farm forestry, individual farmers are being encouraged to plant trees on their own farmland to meet the domestic needs of the family. In many areas this tradition of growing trees on the farmland already exists. Non-commercial farm forestry is the main thrust of most of the social forestry projects in the country today. It is not always necessary that the farmer grows trees for fuel wood, but very often they are interested in growing trees without any economic motive. They may want it to provide shade for the agricultural crops as wind shelters; soil conservation or to use wasteland.

Another scheme taken up under the social forestry programme is, community forestry, which is the raising of trees on community land and not on private land as in farm forestry. All these programmes aim to provide for the entire community and not for any individual. The government has the responsibility of providing seedlings, fertilizer but the community has to take the responsibility of protecting the trees. Some communities manage the plantations sensibly and in a sustainable manner so that the village continues to benefit.

Planting of trees on the sides of roads, canals and railways, along with planting on wastelands is known as 'Extension' Forestry, increasing the boundaries of forests. Under this project there has been creation of wood lots in the village common lands, government wastelands and panchayat lands. Schemes for afforesting degraded government forests that are close to villages are being carried out all over the country.

Hot spot	Original extent of primary vegetation (km^2)	Remaining primary vegetation (km^2)	% of original extent	Area protected (km^2)	% of hotspots	Plant species	Endemic plants	% of global plants (300,000)	Vertebrate species	Endemic vertebrates	% of global vertebrates (27298)
Tropical Andes	1,258,000	314,500	25.0	79,687	25.3	45,000	20,000	6.7	3,389	1,567	5.7
Mesoamerica	1,155,000	231,000	20.0	138,437	59.9	24,000	5,000	1.7	2,859	1,159	4.2
Caribbean	263,500	29,840	11.3	29,840	100.0	12,000	7,000	2.3	1,518	779	2.9
Brazil's Atlantic Forest	1,227,600	91,930	7.5	33,084	35.9	20,000	8,000	2.7	1,361	567	2.1
Choc.Darien/Western Ecuador	260,600	63,000	24.2	16,471	26.1	9,000	2,250	0.8	1,625	418	1.5
Brazil's Cerrado	1,783,200	356,630	20.0	22,000	6.2	10,000	4,400	1.5	1,268	117	0.4
Central Chile	300,000	90,000	30.0	9,167	10.2	3,429	1,605	0.5	335	61	0.2
California Floristic Province	324,000	80,000	24.7	31,443	39.3	4,426	2,125	0.7	584	71	0.3
Madagascar [a]	594,150	59,038	9.9	11,548	19.6	12,000	9,704	3.2	987	771	2.8
Eastern Arc and Coastal Forests of Tanzania/Kenya	30,000	2,000	6.7	2,000	100.0	4,000	1,500	0.5	1,019	121	0.4
Western African Forests	1,265,000	126,500	10.0	20,324	16.1	9,000	2,250	0.8	1,320	270	1
Cape Floristic Province	74,000	18,000	24.3	14,060	78.1	8,200	5,682	1.9	562	53	0.2

Hot spot	Original extent of primary vegetation (km^2)	Remaining primary vegetation (km^2)	% of original extent	Area protected (km^2)	% of hotspots	Plant species	Endemic plants	% of global plants (300,000)	Vertebrate species	Endemic vertebrates	% of global vertebrates (27298)
Succulent Karoo	112,000	30,000	26.8	2,352	7.8	4,849	1,940	0.6	472	45	0.2
Mediterranean Basin	2,362,000	110,000	4.7	42,123	38.3	25,000	13,000	4.3	770	235	0.9
Caucasus	500,000	50,000	10.0	14,050	28.1	6,300	1,600	0.5	632	59	0.2
Sundaland	1,600,000	125,000	7.8	90,000	72.0	25,000	15,000	5.0	1,800	701	2.6
Wallacea	347,000	52,020	15.0	20,415	39.2	10,000	1,500	0.5	1,142	529	1.9
Philippines	300,800	9,023	3.0	3,910	43.3	7,620	5,832	1.9	1,093	518	1.9
Indo-Burma	2,060,000	100,000	4.9	100,000	100.0	13,500	7,000	2.3	2,185	528	1.9
South Central China	800,000	64,000	8.0	16,562	25.9	12,000	3,500	1.2	1,141	178	0.7
Western Ghats/ Sri Lanka	182,500	12,450	6.8	12,450	100.0	4,780	2,180	0.7	1,073	355	1.3
SW Australia	309,850	33,336	10.8	33,336	100.0	5,469	4,331	1.4	456	100	0.4
New Caledonia	18,600	5,200	28.0	526.7	10.1	3,332	2,551	0.9	190	84	0.3
New Zealand	270,500	59,400	22.0	52,068	87.7	2,300	1,865	0.6	217	136	0.5
Polynesia/Micronesia	46,000	10,024	21.8	4,913	49.0	6,557	3,334	1.1	342	223	0.8
Total	17,444,300	2,122,891	12.2	800,767	37.7	[b]	133,149	44.0	[b]	9,645	35

Table 4.2 Biosphere reserves (after Subramanian 2002).

Name of the site	Date of Notification	Location(state)
Nilgiri	1.8.86 Part of	Wyanad, Nagarhole, Bandipur and Madumalai, Nilambur, silent valley and siruvani hills (Tamil nadu, kerala and Karnataka)
Nanda Devi	18.1.88	Part of Chamoli, Pithorgarh & Almora districts(Uttar Pradesh).
Nokerek	01.9.88	Part of Garo Hills (Meghalaya)
Manas	14.3.89	Part of Kokrajhar, Bongaigaon, Barpeta, nalbari, kampruip and Darag District (Assam)
Sunderbans	29.3.89	Part of delta Ganges & Brahmaputra river system (west Bengal)
Gulf Nicobar	06.1.89	Southern most islands of Andaman and Nicobar (A&N Island)
Similipal	21.6.94	Part of Dibrugarh and Tinsukia Districts (Assam)
Dehang Debang	02.09.98	Part of Siang and Debang valley in Arunacahal Pradesh.
Pachmarhi	03.03.99	Part of Betul, Hoshangabad and Chindwara district of Madhya Pradesh
Kanchanjunga	07.2.2000	Part of Kanchanjanga Hills and Sikkim.

Table 4.3 National parks and sanctuaries in India.

Andhra Pradesh	:	Kawal, Pocharam, Parkal Neelapattu, KBR park
Arunachal Pradesh	:	Namidapha
Assam	:	Kaziranga, Manas
Jharkhand	:	Hazaribarh, Betla
Goa	:	Mollen
Gujarat	:	Gir, Velavadar, Wild ass, Nal Sarovar
Haryana	:	Sultanpur lake
H.P.	:	Sechu-tumNallah, Gobind sagar

Table 4.3 ***Contd....***

J & K	:	Dachigam
Karnataka	:	Bandipur, Nagarhole, Rangan Thitoo
M.P.	:	Kanha, Shivpuri,
Maharastra	:	Tadoba, pench, Nawegaon, Bori Dhakna-Kolkaz
Manipur	:	Keibul Lamjao
Meghalaya	:	Balpakrani
Mizoram	:	Dampa
Nagaland	:	Intangki
Orissa	:	Simplipal, Satkasia, Chilka lake
Punjab	:	Abohar
Rajastan	:	Ranthambore, Ghana, Sariska, Bandhavgarth
Sikkim	:	Kanchenjuga
Tamil Nadu	:	Guindy, Mudumalai, Annamalai, Vedanthangal, Vettangudi.
U.P.	:	Corbett, Dudwa
West Bengal	:	Sajnakhali, Mahanandi, Jaldapara, Deer Parks, Sunderban.

REVIEW QUESTIONS

1. Define Biodiversity and explain various types of Biodiversity.
2. Discuss briefly about the threats to biodiversity.
3. Explain Hot spots of Biodiversity in India.
4. List out National Parks and sanctuaries in India.
5. Explain the Biodiversity conservation procedure.
6. Write an essay on values of Biodiversity.
7. Write short notes on (i) Major strategies of biodiversity conservation (ii) Uses of forests (iii) Biosphere reserves.

CHAPTER 5

ENVIRONMENTAL POLLUTION

5.1 CLASSIFICATION OF POLLUTION AND POLLUTANTS

Pollution is defined as the presence of impurities or pollutant substances in sufficient concentration levels, causing harmful effects on human beings, animals, plant life or material resources when exposed for a sufficient duration of time, thus reducing the quality of life in the environment. Pollution is the effect of undesirable changes in our surroundings that have harmful effects on plants, animals and human beings. Pollutants include solid, liquid or gaseous substances present in greater than natural abundance, produce due to human activity, which have a determined effect on our environment. The nature and concentration of a pollutant determine the severity of its detrimental effects on human health. Impurities released directly from the source of origin are known as primary pollutants, for example, CO, SO_2, NO. When contaminants like HC, NO, O_3, combine in the atmosphere (moisture, sunlight) to form new products like PAN (peroxy acetyl nitrate), petrochemical smog, formaldehyde, which are known as secondary pollutants. From an ecological perspective, pollutants can be classified as degradable, slowly degradable and non-degradable. The Degradable or called as non-persistent pollutants can be rapidly broken down by natural process. For example domestic sewage, discarded vegetables, Slowly-degradable or persistent pollutants are pollutants that remain in the environment for many years in an unchanged condition and take decades or longer to degrade. For example, DDT (pesticides) and most plastics. The Non-degradable pollutants cannot be degraded by natural process. Once they are released into the environment they are difficult to eradicate and continue to accumulate: For example, toxic elements like lead or mercury, and nuclear wastes.

5.2 CAUSES, EFFECTS AND SOURCES

5.2.1 Causes

The ultimate cause of pollution is human activity itself. Pollution is a human contribution to nature. Human activities mainly include: industries for various human needs both directly and indirectly, agriculture for food production and industrial needs, health care for health of human beings and animals, transport for mobility of human beings, dwelling for settlement in city or villages, energy for various direct human and industrial needs.

All of these activities contribute to pollution in one way or other and therefore causes miseries. All of them are aimed to be part of human welfare programmes. Along with welfare, all of them have brought the maladies of pollution. A vast array of industries can cause pollution contrary to popular perception that only a chemical industry can cause pollution. The nature and intensity of pollution may be different in different industry. In others, it may be invisible, indirect or negligible. In such a broad sense, no industry is free of pollution. Table 5.1 shows the Classified list of industries causing different types of pollution.

Table 5.1 List of Industries causing different types of pollution.

Type of industries	Type of pollution
Manufacture of chemicals, pesticides, medicines	Water pollution, air pollution
Manufacture of gases	Air pollution
Cement, steel and other mine based industries	Air pollution and solid wastes, noise pollution
Textile industries and their ancillaries	Water pollution, air pollution, noise pollution
Transport vehicle manufacturing	Solid wastes, noise pollution, air pollution
Petroleum based industries	Water pollution, air pollution
Forest dependent industries	Air pollution, solid wastes and sound pollution
Food industries	Water pollution, air pollution, food pollution
Paper industries	Water pollution, air pollution, solid wastes, sound pollution
Sugar industry	Water pollution, air pollution, solid wastes
Brick industry	Air pollution, water pollution
Aircraft industry	Solid wastes, water pollution, air pollution
Electrical appliances and electric goods industries	Solid wastes, air pollution
IT based industries	Air pollution
Telecom industries	Solid wastes and air pollution

5.2.2 Effects of Pollution

Pollution produces physical and biological effects that vary from mildly irritating to lethal. The more serious of the two are the biological effects. The physical effects of pollution are those that we can see, but they include effects other than actual physical damage. Oils spills, killing fish birds, coal produce sox, particulates etc., are considered as physical effects. Air pollutants speed the erosion of statues and buildings, which in some instances, destroys works of art. The most serious result of pollution is its harmful biological effects on human health and on the foodchain of animals, birds, and marine life. Pollution can destroy vegetation that provides food and shelter. It disrupt the balance of nature, and, in extreme cases, can cause the death of humans. Pesticides, which include herbicides and insecticides, can damage crops; kill vegetation; and poison birds, animals, and fish. Most pesticides are nonselective; they kill or damage life forms other than those intended For example, pesticides used in an effort to control or destroy undesirable vegetation and insects often destroy birds and small animals. The biological effect of water pollution is its danger to our water supplies. Water pollutants are also dangerous to all forms of marine life.

5.2.3 Sources of Pollution

Nature contributes to pollution by eroding the soil casing silt to build up in streams and by volcanic eruptions that pollute the atmosphere. However, people cause most pollution problems in the world. The main sources of pollutants are agricultural, industrial, municipal and transportation operations. Agricultural pollutants include insecticides, herbicides, pesticides, natural and chemical fertilizers, drainage from animal feedlots, salts from field irrigation, and silts from uncontrolled soil erosion. Industrial operations produce a wide variety of pollutants. Industrial pollutants include acids from mines and factories, thermal discharges from power plants and radioactive wastes from mining and processing certain ores. Industries create pollutants by producing food, chemicals, metals, petroleum products, and poisons, as well as countless other by-products of our country's technology. The primary municipal pollutants are raw or inadequately treated sewage. Other municipal pollutants include refuse, storm-water overflows and salts used on streets in wintertime. Transportation pollutants include emissions from aircraft, trains, waterborne vessels, and cars and trucks. Motor vehicles create most of our air pollutants through their release of unburned fuel vapors (hydrocarbons).

5.3 AIR POLLUTION

The atmosphere is normally composed of 79% nitrogen, 20% oxygen and 1% as a mixture of carbon dioxide, water vapor and trace amounts of several other gases such as neon, helium, methane, krypton, hydrogen and xenon. The general structure of the atmosphere has several important features that have relevance to environmental problems.

Air pollution occurs due to the presence of undesirable solid or gaseous particles in the air, in quantities that are harmful to human health and the environment. The air may become polluted by natural causes such as Volcanos, which release ash, dust, sulfur and other gases or by forest fires that are occasionally naturally causes by lightening. However unlike pollutants from human activity naturally occurring pollutants tend to remain in the atmosphere for a short time and do not lead to permanent atmospheric change.

5.3.1 Primary and Secondary Pollutants

(a) Pollutants that are emitted directly from identifiable sources are produced both by natural events (e.g., dust storms and volcanic eruptions) and human activities (emission from vehicles, industries, etc.). These are called primary pollutants. There are five primary pollutants that together contribute to about 90% of the global air pollution. These are carbon oxides (CO & CO2), nitrogen oxides, sulfur oxides, volatile organic compounds (mostly hydrocarbons), and suspended, particulate matter. The pollutants that are produced in the atmosphere when certain chemical reactions take place among the primary pollutants are called secondary pollutants; e.g., sulfuric acid, nitric acid, carbonic acid, etc. Sources of air pollution can be broadly classified into natural and anthropogenic. The natural sources include volcano, forest fire and pollens. The anthropogenic sources include every thing involving human activities.

(b) The atmosphere is divided into several layers. Air pollution occurs due to the presence of undesirable solid or gaseous particles in the air, in quantities that are harmful to human health and the environment. The air may become polluted by natural causes such as volcanoes, which release ash, dust, sulfur and other gases, or by forest fires that are occasionally naturally caused by lightening. However, unlike pollutants from human activity, naturally occurring pollutants tend to remain in the atmosphere for a short time and do not lead to permanent atmospheric change.

5.3.2 Automobile Pollution

During the last decade the automobile pollution has increased manifold in Metropolitan cities like Kolkata, Mumbai, Chennai, Hyderabad. Automobile exhausts emit particulate matter, unburnt hydrocarbons, carbon monoxide, nitrogen oxides and some carcinogenic organic compounds like benzene and polyaromatic hydrocarbons, having adverse health impact among the exposed population. Various other factors like predominance of old outdated vehicles, lack of maintenance, limited available road area, poor traffic management and poor road conditions helped accentuate the level of automobile pollution. Rapid increase in the number of vehicles is another reason for increasing automobile pollution. Inspection and maintenance of vehicles is an important component for controlling emission from in-use vehicles irrespective of the state of- engine technology and its after-treatment system. Tail pipe emissions from the vehicles are monitored to check compliance of regulatory standard in this regard. The Auto Emission Testing Centers (AETCs) play an important role in monitoring emission from in-use vehicles. These centres issue Pollution- Under-Control (PUC) certificate if the vehicles are found to emit exhaust gases complying with the statutory emission standards. But most of these centres do not have adequate equipment for testing. Either the equipment up gradation of Auto Emission Testing Centres is not calibrated or the centre lacks adequate qualified manpower.

5.3.3 Industrial Pollution

Industrial pollution is pollution which can be directly linked with industry, in contrast to other pollution sources. This form of pollution is one of the leading causes of pollution worldwide. In India, for example, the Ministry of Environment and Forests estimates that up to 50% of the nation's pollution is caused by industry. Because of its size and scope, industrial pollution is a serious problem for the entire planet, especially in nations which are rapidly industrializing, like China and India. The Industrial Revolution mechanized means of production, allowing for a much greater volume of production, and generating a corresponding increase in pollution. The problem was compounded by the use of fuels like coal, which is notoriously unclean, and a poor understanding of the causes and consequences of pollution. There are a number of forms of industrial pollution. One of the most common is water pollution, caused by dumping of industrial waste into waterways, or improper containment of waste, which causes leakage into groundwater and waterways. Industrial pollution can also impact air quality, and it can enter the soil, causing widespread environmental problems. Because of the nature of the global environment, industrial pollution is never limited to industrial nations. Samples of ice cores from Antarctica and the Arctic both show high levels of industrial pollutants, illustrating the immense distances which pollutants can travel, and traces of industrial pollutants have been identified in isolated human, animal, and plant populations as well.

Industrial pollution hurts the environment in a range of ways, and it has a negative impact on human lives and health. Pollutants can kill animals and plants, imbalance ecosystems, degrade air quality radically, damage buildings, and generally degrade quality of life. Factory workers in areas with uncontrolled industrial pollution are especially vulnerable. A growing awareness of factory pollution and its consequences has led to tighter restrictions on pollution all over the world, with nations recognizing that they have an obligation to protect themselves and their neighbors from pollution. However, industrial pollution also highlights a growing issue: the desire of developing nations to achieve first world standards of living and production. As these countries industrialize, they add to the global burden of industrial pollution, triggering serious discussions and arguments about environmental responsibility and a desire to reach a global agreement on pollution issues. Industrial pollution is clearly one of the biggest contributors to our polluted landscapes.

5.3.4 Ambient Air Quality Standards

The ambient air Quality standards given by CPCB are given in Table 5.2.

Table 5.2 National Ambient Air Quality Standards.

S.No	Pollutant	Time Weighted Average	Concentration in ambient air		
			Industrial, Residential, Rural and Other Area	Ecologically Sensitive Area (notified by Central Government)	Method of Measurement
(1)	(2)	(3)	(4)	(5)	(6)
1	Sulphur Dioxide (SO_2), μg/m^3	Annual* 24 hours**	50 80	20 80	- Improved West and Gaeke - Ultraviolet fluorescence
2	Nitrogen Dioxide (NO_2), μg/m^3	Annual* 24 hours**	40 80	30 80	- Modified jacob & Hochhelser (Na-Arsenlte) - Chemiluminescence
3	Particulate Matter (size less than 10mm) or PM_{10} μg/m^3	Annual* 24 hours**	60 100	60 100	- Gravimetric - TOEM - Beta attenuation

Table 5.2 *Contd....*

(1)	(2)	(3)	(4)	(5)	(6)
4	Particulate Matter (size less than 2.5μm) or $PM_{2.5}$ μg/m^3	Annual* 24 hours**	40 60	40 60	- Gravimetric - TOEM - Beta attenuation
5	Ozone (O_3) μg/m^3	8 hours** 1 hour**	100 180	100 180	- UV photometric - Chemilminescence - Chemical Method
6	Lead (Pb) μg/m^2	Annual* 24 hours**	0.50 1.0	0.50 1.0	- AAS/ICP method after sampling on EPM 2000 or equivalent filter paper - ED-XRF using Teflon filter
7	Carbon monoxide (CO) mg/m^3	8 hours** 1 hour**	02 04	02 04	- Non Dispersive Infra Red (NDIR) spectroscopy
8	Ammonia (NH_3) μg/m^3	Annaul* 24 hours**	100 400	100 400	- Chemiluminescence - Indophenol blue method
9	Benzene (C_6H_4) μg/m^3	Annual*	05	05	- Gas chromatography based continuous analyzer - Adsorption and Desorption followed by GC analysis
10	Benzo(a)Pyrene (BaP) – particulate phase only, ng/m^3	Annual*	06	06	- AAS/ICP method after sampling on EPM 2000 or equivalent filter paper
11	Arsonic (As), ng/m^3	Annual*	06	06	- AAS/ICP method after sampling on EPM 2000 or equivalent filter paper
12	Nickel (Ni), ng/m^3	Annual*	20	20	- AAS/ICP method after sampling on EPM 2000 or equivalent filter paper

* Annual arithmetic mean of minimum 104 measurements in a year at a particular site taken twice a week 24 hourly at uniform intervals.

** 24 hourly or 08 hourly or 01 hourly monitored alues, as applicable, shall be compiled with 98% of the time in year. 2% of the time, they may exceed the limits but not on two consecutive days of monitoring.

Note: Whenever and wherever monitoring results on two consecutive days of monitoring exceed the limits specified above for the respective category, it shall be considered adequate reason to institute regular or continuous monitoring and further investigation.

NOTE:

1. *National Ambient Air Quality Standard:* The levels of air quality with an adequate margin of safety, to protect the public health, vegetation and property.
2. Whenever and wherever two consecutive values exceeds the limit specified above for the respective category, it would be considered adequate reason to institute regular / continuous monitoring and further investigations.

5.4 WATER POLLUTION

Although 71% of the Earth's surface is covered by water, only a tiny fraction of the water is available to us as fresh water. About 97% of the total water available on Earth is found in the oceans and is too salty for drinking or irrigation. The remaining 3% is fresh water. Of this, 2.997% is locked in ice caps or glaciers. Thus, only 0.003% of the Earth's total volume of water is easily available to us as soil moisture, ground water, water vapor and the water in lakes, streams, rivers and wetlands.

In short, if the world's water supply were only 100 liters our usable supply of freshwater would be only about 0.0003 liters (one-half teaspoon). This makes water a very precious resource. The future wars in our world may well be fought over water. By the middle of this century, almost twice as many people will be trying to share the same amount of fresh water the Earth has today. In the future, as fresh water becomes scarcer, the access to water resources will be an important factor in determining the economic growth of several countries around the world. When the quality or composition of water changes directly or indirectly as a result of men's activities such that it becomes unfit for any purpose it is said to polluted. When a source of pollution can be readily identified because it has a definite source and place where it enters the water it is said to come from a point source, e.g., municipal and industrial discharge pipes. When a source of pollution cannot be readily identified, such as agricultural runoff, acid rain, etc., they are said to be non-point sources of pollution.

5.4.1 Point and Non Point Source of Pollution

Broadly, water pollution causes due to either point source or non point source or both - the Point sources are like drain pipes, ditches, sewer out falls etc., and Non-Point Sources are runoff, atmosphere deposition etc.

Water pollution may be caused by biotic and a biotic contaminants. The water pollution results form either natural sources or artificial sources (i.e., human activities). The natural phenomena, for example, drought renders water levels to decline in aquifers allowing salt water to intrude; during rains, the running water gathers silt, harmful chemicals such as fertilizers and pesticides from the soil and so on. Sewage is another major cause of water pollution. The human activities, industry, agriculture, mining, deforestation, power generation, cause water pollution. Population growth and unplanned urbanization also contribute to pollution of water. Factories and mines release large quantities of toxic chemicals, organic wastes, heavy metals, heated effluents, inert wastes and radioactive wastes causing water pollution. Toxic chemicals / wastes are not easily degradable by biological means. DDT and mercury fall under this category. The water contaminated by them is highly poisonous and if contacted or consumed by plants or animals, may prove fatal. Other examples are pesticides and herbicides which wash off from the land during rainy season into the water sources. Organic wastes refer to rotting organic matter, which generate foul smell. Examples are waste from humans, tanneries, and paper and pulp industries.

These wastes, when discharged into water sources decompose using large amounts of oxygen from water rendering water useless. The pollution level is indicated by a parameter, BOD (bio-chemical oxygen demand). With oxygen depletion, some types of fish perish in these water bodies. Urban sewage is another example of organic waste, generally released into rivers, lakes or tanks in most of the developing countries. The inert wastes such as metals, oil films, silt settle to the bottom, if not removed, and affects plant life at the bottom of the water bodies. This is not a very serious pollution problem. The radioactive wastes are generally dumped into oceans; if proper disposal techniques are not followed, these wastes create serious pollution of oceans. This aspect is discussed later. It is estimated that many countries will have fresh water shortage by 2025. Due to water pollution two thirds of world population may live in water stressed conditions. Depletion in groundwater has been causing alarming water is wasted. Dams may ensure a year round water supply, but much water gets evaporated and experiences loss due to seepage. In summary, industrial units, urban centers and farming activities are the principal sources of water pollution.

5.4.2 Major Pollutants of Water and their Sources

The following are the common types of water pollutants:

Disease-causing agents: Bacteria, viruses, protozoa and parasitic worms that enter water from domestic sewage and animal wastes. In developing countries they are the major cause of sickness and death, prematurely killing an average of 13,700 people each day.

Oxygen-demanding wastes: Organic wastes, which can be decomposed by oxygen-consuming bacteria. Large populations of bacteria supported by these wastes can deplete water of dissolved oxygen gas. Without enough oxygen, fish and other oxygen consuming forms of aquatic life die.

Water-soluble inorganic chemicals: Acids, salts, and compounds of toxic metals such as lead and mercury. Such dissolved solids can make water unfit to drink, harm fish and other aquatic life, decrease crop yields, and accelerate corrosion of equipment that uses of water.

Inorganic plant nutrients: Water soluble nitrate and phosphate compounds that can cause excessive growth of algae and other aquatic plants, which than die and decay, depleting water of dissolved oxygen and killing fish.

Organic chemicals: Oil, gasoline, plastics, pesticides, cleaning solvents, detergents and many other water soluble and insoluble chemicals that threaten human health and harm fish and other aquatic life.

Sediment or suspended matter: Insoluble particles of soil and other solid inorganic and organic materials that become suspended in water and that in turns of total mass are the largest source of water pollution. Suspended particulate matter clouds the water, reduces the ability of some organisms to find food, reduces photosynthesis by aquatic plants, disrupts aquatic food webs, and carries pesticides, bacteria and other harmful substances. Bottom sediments destroys feeding and spawning grounds of fish and clogs and fills lakes, artificial reservoirs, stream channels and harbors.

Radioactive substances : Radioisotopes that are water soluble or capable of being biologically amplified to higher concentrations as they pass through food chains and webs. Ionizing radiation from such isotopes can cause birth defects, cancer, and genetic damage.

Heat: Large quantity of water is heated when it is used in the cooling towers of thermal power plants. When this hot water is discharged in to the near by water bodies it causes an increase in its temperature. This increase in water temperature lowers dissolved oxygen content and makes aquatic organisms more vulnerable to disease, parasites and toxic chemicals.

5.4.3 Water Quality Requirement for Different Uses

For any water body to function adequately in satisfying the desired use, it must have corresponding degree of purity. Drinking water should be of highest purity. As the magnitude of demand for water is fast approaching the available supply, the concept of management of the quality of water is becoming as important as its quantity.

Each water use has specific quality need. Therefore, to set the standard for the desire quality of a water body, it is essential to identify the uses of water in that water body. In India, the Central Pollution Control Board (CPCB) has developed a concept of *designated best use*. According to this, out of the several uses of water of a particular body, the use which demands highest quality is termed its *designated best use*. Five *designated best uses* have been identified. This classification helps the water quality managers and planners to set water quality targets and design suitable restoration programs for various water bodies. Table 5.3 to Table 5.7 shows the water quality standards for various uses.

Table 5.3

Designated Best Use	Class	Criteria
Drinking Water Source without conventional treatment but after disinfection	A	1. Total Coliforms Organism MPN/100ml shall be 50 or less 2. pH between 6.5 and 8.5 3. Dissolved Oxygen 6mg/l or more 4. Biochemical Oxygen Demand 5 days 20 °C, 2mg/l or less
Outdoor bathing (Organised)	B	1.Total Coliforms Organism MPN/100ml shall be 500 or less 2. pH between 6.5 and 8.5 3. Dissolved Oxygen 5mg/l or more 4. Biochemical Oxygen Demand 5 days 20 °C, 3mg/l or less
Drinking water source after conventional treatment and disinfection	C	1. Total Coliforms Organism MPN/100ml shall be 5000 or less 2. pH between 6 and 9 3. Dissolved Oxygen 4mg/l or more 4. Biochemical Oxygen Demand 5 days 20 °C, 3mg/l or less
Propagation of Wild life and Fisheries	D	1. pH between 6.5 and 8.5 2. Dissolved Oxygen 4mg/l or more 3. Free Ammonia (as N) 4. Biochemical Oxygen Demand 5 days 20 °C, 2mg/l or less
Irrigation, Industrial Cooling, Controlled Waste disposal	E	1. pH between 6.0 and 8.5 2. Electrical Conductivity at 25 °C micro mhos/cm, maximum 2250 3. Sodium absorption Ratio Max. 26 4. Boron Max. 2mg/l
	Below-E	Not meeting any of the A, B, C, D & E criteria

Table 5.4 A colour coding frequently used to depict the quality of water on maps

Blue water	This water can be directly used for drinking, industrial use, etc.
Green water	Water contained in soil and plants is termed as green water
White water	Atmospheric moisture is white water
Brown or grey water	Various grades of wastewater are shown by brown or grey colour

In India, CPCB has identified water quality requirements in terms of a few chemical characteristics, known as primary water quality criteria. Further, Bureau of Indian Standards has also recommended water quality parameters for different uses in the standard IS 2296:1992.

Table 5.5 Water Quality Standards in India

Characteristics	**Designated best use**				
	A	**B**	**C**	**D**	**E**
Dissolved Oxygen (DO)mg/l, min	6	5	4	4	-
Biochemical Oxygen demand (BOD)mg/l, max	2	3	3	-	-
Total coliform organisms MPN/100ml, max	50	500	5,000	-	-
pH value	6.5-8.5	6.5-8.5	6.0-9.0	6.5-8.5	6.0-8.5
Colour, Hazen units, max.	10	300	300	-	-
Odour	Un-objectionable			-	-
Taste	Tasteless	-	-	-	-
Total dissolved solids, mg/l, max.	500	-	1,500	-	2,100
Total hardness (as $CaCO_3$), mg/l, max.	200	-	-	-	-
Calcium hardness (as $CaCO_3$), mg/l, max.	200	-	-	-	-
Magnesium hardness (as $CaCO_3$), mg/l, max.	200	-	-	-	-
Copper (as Cu), mg/l, max.	1.5	-	1.5	-	-
Iron (as Fe), mg/l, max.	0.3	-	0.5	-	-
Manganese (as Mn), mg/l, max.	0.5	-	-	-	-
Cholorides (as Cu), mg/l, max.	250	-	600	-	600
Sulphates (as SO_4), mg/l, max.	400	-	400	-	1,000
Nitrates (as NO_3), mg/l, max.	20	-	50	-	-
Fluorides (as F), mg/l, max.	1.5	1.5	1.5	-	-
Phenolic compounds (as C_2H_5OH), mg/l, max.	0.002	0.005	0.005	-	-
Mercury (as Hg), mg/l, max.	0.001	-	-	-	-
Cadmium (as Cd), mg/l, max.	0.01	-	0.01	-	-
Salenium (as Se), mg/l, max.	0.01	-	0.05	-	-
Arsenic (as As), mg/l, max.	0.05	0.2	0.2		

(Source IS 2296:1992)

Table 5.6 Indian Standard Specifications for Drinking Water is: 10500

S.No.	Parameter	Requirement desirable Limit	Remarks
1	Colour	5	May be extended up to 50 if toxic substances are suspected
2.	Turbidity	10	May be relaxed up to 25 in the absence of alternate
3.	pH	6.5 to 8.5	May be relaxed up to 9.2 in the absence
4.	Total Hardness	300	May be extended up to 600
5.	Calcium as Ca	75	May be extended up to 200
6.	Magnesium as Mg	30	May be extended up to 100
7.	Copper as Cu	0.05	May be relaxed up to 1.5
8.	Iron	0.3	May be extended up to 1
9.	Manganese	0.1	May be extended up to 0.5
10.	Chlorides	250	May be extended up to 1000
11.	Sulphates	150	May be extended up to 400
12.	Nitrates	45	No relaxation
13.	Fluoride	0.6 to 1.2	If the limit is below 0.6 water should be rejected, Max. Limit is extended to 1.5
14.	Phenols	0.001	May be relaxed up to 0.002
15.	Mercury	0.001	No relaxation
16.	Cadmium	0.01	No relaxation
17.	Selenium	0.01	No relaxation
18.	Arsenic	0.05	No relaxation
19.	Cyanide	0.05	No relaxation
20.	Lead	0.1	No relaxation
21.	Zinc	5.0	May be extended up to 10.0
22.	Anionic detergents (MBAS)	0.2	May be relaxed up to 1
23.	Chromium as Cr+6	0.05	No relaxation
24.	Poly nuclear aromatic Hydrocarbons		
25.	Mineral Oil	0.01	May be relaxed up to 0.03
26.	Residual free Chlorine	0.2	Applicable only when water is chlorinated
27.	Pesticides	Absent	
28.	Radio active		

Table 5.7 Drinking Water Specification: Is: 10500, 1992 (Reaffirmed 1993) Tolerance Limits

S.No	Parameter	IS: 10500 Requirement (Desirable limit)	Undesirable effect outside the desirable limit	IS: 10500 Permissible limit in the absence of alternate source
Essential Characteristics				
1.	pH	6.5 – 8.5	Beyond this range the water will effect the mucous membrane and / or water supply system	No relaxation
2	Colour (Hazen Units), Maximum	5	Above 5, consumer acceptance decreases	25
3	Odour	Unobjectionable		
4.	Taste	Agreeable		
5	Turbidity, NTU, Max	5	Above 5, consumer acceptance decreases	10
Following Results are expressed in mg/1 :				
6.	Total hardness as CaCO3, Max	300	Encrustation in water supply structure and adverse effects on domestic use	600
7.	Iron as Fe, Max	0.30	Beyond this limit taste/appearance are affected, has adverse effect on domestic uses and water supply structures, and promotes iron bacteria.	1.0
8.	Chlorides as Cl, Max	250	Beyond this limit tast, corrosion and palatability are effected	1000
9.	Residual, Free Chlorine, Min	0.20	Beyond this palatability	
Desirable Characteristics				
10.	Dissolved solids, Max	500	decreases and may cause gastro intentional irritation	2000
11.	Calcium as Ca, Max	75	Encrustation in water supply structure and adverse effects on domestic use	200
12.	Magnesium as Mg, Max	30	--	100

Table 5.7 *Contd.....*

13.	Copper as Cu, Max	0.05	Astringent taste, discoloration and corrosion of pipes, fitting and utensils will be caused beyond this	1.5
14.	Manganese as Mn, Max	0.1	Beyond this limit taste/appearance are affected, has adverse effect on domestic uses and water supply structures	0.3
15.	Sulphate as SO4 Max	200	Beyond this causes gastro intentional irritation when magnesium or sodium are present	400
16.	Nitrates as NO3	45	Beyond this methanemoglobinemia takes place	100
17.	Fluoride, Max	1.0	Fluoride may be kept as low as possible. High fluoride may cause fluorosis	1.5
18.	Phenolic compounds as C6H5OH, Max	0.001	Beyond this, it may cause objectionable taste and odour	0.002
19.	Mercury as Hg, Max	0.001	Beyond this, the water becomes toxic	No relaxation
20.	Cadmium as Cd, Max	0.01	Beyond this, the water becomes toxic	No relaxation
21.	Selenium as Se, Max	0.01	Beyond this, the water becomes toxic	No relaxation
22.	Arsenic as As, Max	0.05	Beyond this, the water becomes toxic	No relaxation
23.	Cyanide as CN, Max	0.05	Beyond this, the water becomes toxic	No relaxation
24.	Lead as Pb, Max	0.05	Beyond this, the water becomes toxic	No relaxation
25.	Zinc as Zn, Max	5	Beyond this limit it can cause astringent taste and an opalescence in water	15
26.	Anionic detergents as MBAS, Max	0.2	Beyond this limit it can cause a light froth in water	1.0
27.	Chromium as Cr6+ , Max	0.05	May be carcinogenic above this limit	No relaxation
28.	Ploynuclear aromatic hydrocarbons as PAH, Max	--	May be carcinogenic	--

Table 5.7 *Contd.....*

29.	Mineral Oil, Max	0.01	Beyond this limit undesirable taste and odour after chlorination take place	0.03
30.	Pesticides, Max	Absent	Toxic	0.001
31.	Radioactive materials a) α emitters Bq/1, Max b) β emitters Pci/1, Max	----	----	0.1 1
32.	Alkalinity, Max	200	Beyond this limit taste becomes unpleasant	600
33.	Aluminum as Al, Max	0.03	Cumulative effect is reported to cause dementia	0.2
34.	Boron, Max	1	--	5

5.4.4 Wastewater Treatment Methods

It is estimated that every year 1.8 million people die due to suffering from waterborne diseases. A large part of these deaths can be indirectly attributed to improper sanitation. Wastewater treatment is an important initiative which has to be taken more seriously for the betterment of the society and our future. Wastewater treatment is a process, wherein the contaminants are removed from wastewater as well as household sewage, to produce waste stream or solid waste suitable for discharge or reuse. Wastewater treatment methods are categorized into three sub-divisions, physical, chemical and biological.

5.4.4.1 *Effluent Treatment Plants (ETP)*

Effluent Treatment Plants or (ETPs) are used by leading companies in the pharmaceutical and chemical industry to purify water and remove any toxic and non toxic materials or chemicals from it. These plants are used by all companies for environment protection. An ETP is a plant where the treatment of industrial effluents and waste waters is done. The ETP plants are used widely in industrial sector, for example, pharmaceutical industry, to remove the effluents from the bulk drugs. During the manufacturing process of drugs, varied effluer and contaminants are produced. The effluent treatment plants are used in the removal of high amount of organics, debris, dirt, grit, pollution, toxic, non toxic materials, polymers etc. from drugs and other medicated stuff. The ETP plants use evaporation and drying methods, and other auxiliary techniques such as centrifuging, filtration, incineration for chemical processing and effluent treatment.

The treatment of effluents is essential to prevent pollution of the receiving water. The effluent water treatment plants are installed to reduce the possibility of pollution; biodegradable organics if left unsolved, the levels of contamination in the process of purification could damage bacterial treatment beds and lead to pollution of controlled waters. The ETPs can be established in the industrial sectors like Pharmaceuticals, Chemicals and Leather industry and tanneries.

5.4.4.2 *Sewage Treatment Plants (STP)*

Sewage treatment, or **domestic wastewater treatment**, is the process of removing contaminants from wastewater and household sewage, both runoff (effluents) and domestic. It includes physical, chemical, and biological processes to remove physical, chemical and biological contaminants. Its objective is to produce a waste stream (or treated effluent) and a solid waste or sludge suitable for discharge or reuse back into the environment. This material is often inadvertently contaminated with many toxic organic and inorganic compounds.

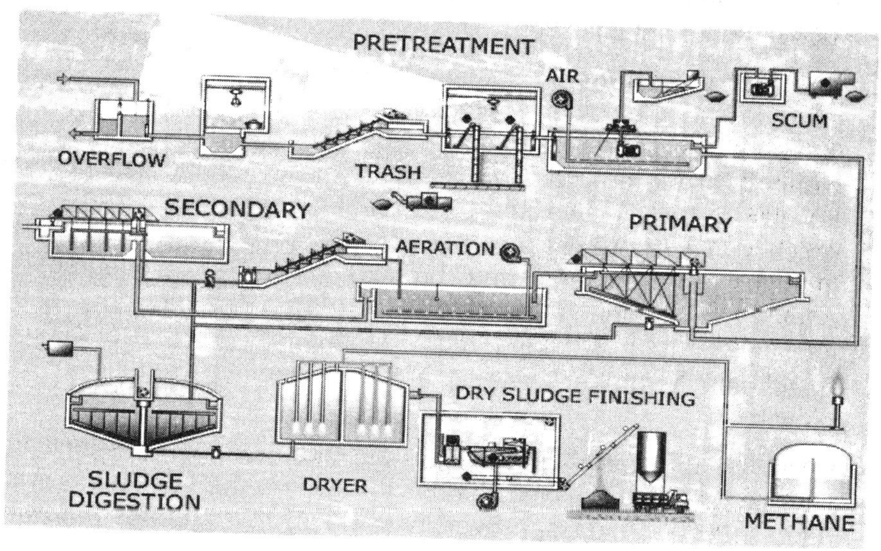

Fig. 5.1 Process Flow Diagram for a typical large-scale treatment plant.

Pre-treatment removes materials that can be easily collected from the raw wastewater before they damage or clog the pumps and skimmers of primary treatment clarifiers, for example, trash, tree limbs, leaves, etc., The influent sewage water is strained to remove all large objects carried in the sewage stream. This is most commonly done with an automated mechanically raked bar screen in modern plants serving large populations, whilst in smaller or less modern plants a manually cleaned screen may be used this is called as screening. The raking action of a mechanical bar screen is typically paced according to the accumulation on the bar screens and/or flow rate. The solids are collected and later disposed in a landfill or incinerated. Pre-treatment may include Grit removal in which, a sand or grit channel or chamber where the velocity of the incoming wastewater is carefully controlled to allow sand, grit and stones to settle.

Primary Treatment: In the primary sedimentation stage, sewage flows through large tanks, commonly called "primary clarifiers" or "primary sedimentation tanks". The tanks are large enough that sludge can settle and floating material such as grease and oils can rise to the surface and be skimmed off. The main purpose of the primary sedimentation stage is to produce both a generally homogeneous liquid capable of being treated biologically and a sludge that can be separately treated or processed. Primary settling tanks are usually equipped with mechanically driven scrapers that continually drive the collected sludge towards a hopper in the base of the tank from where it can be pumped to further sludge treatment stages. Grease and oil from the floating material can sometimes be recovered for specifications.

Secondary Treatment: Secondary treatment is designed to substantially degrade the biological content of the sewage which are derived from human waste, food waste, soaps and detergent. The majority of municipal plants treat the settled sewage liquor using aerobic biological processes. For this to be effective, the biota require both oxygen and a substrate on which to live. There are a number of ways in which this is done. In all these methods, the bacteria and protozoa consume biodegradable soluble organic contaminants (e.g. sugars, fats, organic short-chain carbon molecules, etc.) and bind much of the less soluble fractions into floc. Secondary treatment systems are classified as fixed-film or Suspended-growth.

Fixed-film or attached growth system treatment process including trickling filter and rotating biological contactors where the biomass grows on media and the sewage passes over its surface. In suspended-growth systems, such as activated sludge, the biomass is well mixed with the sewage and can be operated in a smaller space than fixed-film systems that treat the same amount of water. However, fixed-film systems are more able to cope with drastic changes in the amount of biological material and can provide higher removal rates for organic material and suspended solids than suspended growth systems. Roughing filters are intended to treat particularly strong or variable organic loads, typically industrial, to allow them to then be treated by conventional secondary treatment processes. Characteristics include typically tall, circular filters filled with open synthetic filter media to which wastewater is applied at a relatively high rate. They are designed to allow high hydraulic loading and a high flow-through of air. On larger installations, air is forced through the media using blowers. The resultant wastewater is usually within the normal range for conventional treatment processes.

Activated Sludge: In general, activated sludge plants encompass a variety of mechanisms and processes that use dissolved oxygen to promote the growth of biological floc that substantially removes organic material. The process traps particulate material and can, under ideal conditions, convert ammonia to nitrite and nitrate and ultimately to nitrogen gas.

5.4.4.3 *Common and Combined Effluent Treatment Plants (CETP)*

Many of the Small Scale Industries (SSI) are unable to put up the treatment systems individually, the concept of CETP's (Common Effluent Treatment Plants) is envisaged to benefit such industries in treating its effluent before disposal whether it is in stream, land, sewerage system or in rivers and seas. CETP's are set up in the industrial estates where there are clusters of small scale industrial units and where many polluting industries are located. The Ministry of Environment & Forest, Govt. of India has launched the centrally sponsored scheme, namely, Common Effluent Treatment Plant (CETP) in order to make a cooperative movement of pollution control especially to treat the effluent, emanating from the clusters of compatible Small-Scale Industries. The major objective of the CETP is therefore, to reduce the treatment cost to be borne by an individual member unit to a maximum while protecting the water environment to a maximum. The proposal for setting up of CETP's by such industries is to be submitted by the CETP Association to the respective State Pollution Control Board which after examining the proposal and obtaining commitment from the concerned State Government regarding its contribution will give their recommendation to the Ministry of Environment and Forests for consideration , the Ministry examines the proposal and takes the decision through a Screening Committee constituted in this regard for providing support from the Central Government.

The Ministry releases the funds for the approved projects which is the matching grant to the amount released by the concerned State Government, subject to the bank guarantee to be taken from the CETP associations for the amount released by the Central Government, The CETP Company should meet the remaining cost by equity contribution by the industries and loans from financial institutions. Funds released for the

CETP's should be utilized for the CETP only and not for payment for any debts/bank loans, etc. The design and technical specification of CETP can be referred in any book on wastewater management.

5.5 SOIL POLLUTION

Soil is the thin layer of organic and inorganic material that covers the Earth's rocky surface. The organic portion, which is derived from the decayed remains of plants and animals, is concentrated in the dark uppermost "topsoil". The inorganic portion, which is made up of rock fragments, is formed over thousands of years by physical and chemical weathering of bedrock.

Soil Pollution is the introduction of substances or biological organisms, into the soil, resulting in a change of the soil quality this affects the normal use of the soil or endangering public health and the living environment. Soil contaminants are spilled onto the surface through many different activities.

5.5.1 Impact of Modern Agriculture on Soil

Industrial agriculture is a form of modern farming that refers to the industrialized production of livestock, poultry, fish, and crops. The methods of industrial agriculture are techno scientific, economic, and political. They include innovation in agricultural machinery and farming. Modern methods of agriculture have resulted in use of fertilizers and pesticides to increase the yield of the crops. Most of them are synthetic and chemicals-based. They are collectively called agro-chemicals. When a nucleus emits radiation, it is said to decay. Rays from radioactive isotopes are dangerous to living things but are also useful in variety of ways. Exposure to large amounts of radiation is harmful to health. The Greenhouse effect is the rise in temperature that the Earth experiences because certain gases in the atmosphere referred to as greenhouse gases, trap energy from the Sun. Destruction of wild life leads to loss of valuable materials. 2) Disappearance of any link in a food chain upsets nature's balance and creates problems. 3) New improved varieties of crops and other useful animals are derived from their wild relatives by genetic modification.

5.5.2 Degradation of Soil

Land degradation refers to a decline in the overall quality of soil, water or vegetation condition commonly caused by human activities. This leads to soil erosion, rising water tables, the expression of salinity, mass movement by gravity of soil or rock, stream bank instability and a process that results in declining water quality. Land degradation is a global problem, largely related to agricultural use. The major causes include:

- Land clearance, such as clear cutting and deforestation
- Agricultural depletion of soil nutrients through poor farming practices
- Livestock including overgrazing
- Irrigation and over drafting
- Urban sprawl and commercial development
- Land pollution including industrial waste
- Vehicle off-roading
- Quarrying of stone, sand, ore and minerals

5.6 MARINE POLLUTION

Marine pollution occurs when harmful effects, or potentially harmful effects, can result from the entry into the ocean of chemicals, particles, industrial, agricultural and residential waste, noise, or the spread of invasive organisms. Most sources of marine pollution are land based. The pollution often comes from non point sources such as agricultural runoff and wind blown debris. Many potentially toxic chemicals adhere to tiny particles which are then taken up by plankton and benthos animals, most of which are either deposit or filter feeders. In this way, the toxins are concentrated upward within ocean food chains. Many particles combine chemically in a manner highly depletive of oxygen, causing estuaries to become anoxic. When pesticides are incorporated into the marine ecosystem, they quickly become absorbed into marine food webs. Once in the food webs, these pesticides can cause mutations, as well as diseases, which can be harmful to humans as well as the entire food web. Toxic can also be introduced into marine food webs. These can cause a change to tissue matter, biochemistry, behavior, reproduction, and suppress growth in marine life. Also, many animal feeds have a high fish meal or fish hydrolysate content. In this way, marine toxins can be transferred to land animals, and appear later in meat and dairy products.

5.6.1 Misuse of International Water for Dumping of Hazardous Waste

Ocean dumping is the practice of dumping all sorts of garbage in the oceanic waters. Industrial waste reigns supreme in this process, since the oceans provide a convenient out-of-sight, out-of-mind way to get rid of the junk that industries churn out during the course of production. A report by the United Nations Environmental Programme (UNEP) says that plastic, especially plastic bags and polyethylene terephthalate (PET) is the most common of all ocean pollutants. Also, plastic is the most dangerous to be dumped into the ocean with regards to marine life. Easily mistakable as food by marine mammals, fish, birds and turtles, plastic proves lethal for them. There are three main direct public health risks from ocean dumping: the first one being the risk to humans by the consumption of contaminated marine organisms, the other occupational hazards of injuries and accidents faced by the fishermen whose livelihood depends upon the state of the oceans and the exposure of humans to the waste that is washed up on the beaches. Consumption of fish contaminated from radioactive wastes may pose a serious problem worldwide because of nuclear waste dumping in the oceans.

In India, there is hardly any law that safeguards the cleanliness of its beaches and thus, all the beach coast of the country is free to be littered by the many tourists that visit the glorious waters of the Bay of Bengal, Arabian Sea and the Indian Ocean. To add to that, the Indian government, in 2007 proposed a newly drafted hazardous waste management law that seeks to undo established, science-based definitions of waste and consider waste that is being recycled somehow less hazardous than the waste being land filled in order to curry favor with hazardous scrapping industries.

5.6.2 Coastal Pollution Due to Industrial Effluents

Coastal waters receive a variety of land-based water pollutants, ranging from petroleum wastes to pesticides to excess sediments. Marine waters also receive wastes directly from offshore activities, such as ocean-based dumping, for example, from ships and offshore oil and gas operations. One pollutant in the ocean is sewage. Human sewage largely consists of excrement from toilet-flushing; wastewater from bathing, laundry, and dishwashing; and animal and vegetable matter from food preparation that is disposed through an in-sink garbage disposal. The major types of ocean pollutants from industrial sources can be generally categorized as petroleum, hazardous, thermal, and radioactive. Petroleum products are oil and oil-derived chemicals used for fuel, manufacturing, plastics-making, and many other purposes. Hazardous wastes are chemicals that are toxic (poisonous at certain levels), reactive (capable of producing explosive gases), corrosive (able to corrode steel), or ignitable (flammable). Thermal wastes are heated wastewaters, typically from power plants and factories, where water is used for cooling purposes. Radioactive wastes contain chemical elements having an unstable nucleus that will spontaneously decay with the concurrent emission of ionizing radiation.

Industrial wastes primarily enter coastal waters from terrestrial or land-based activities. Industries, like municipalities and other entities that generate wastes, dispose of many liquid wastes through wastewater systems and ultimately to water bodies, whereas they dispose of their solid wastes in landfills. The quantity and characteristics of industrial wastewater depends on the type of industry, its water and wastewater management, and its type of waste pretreatment before delivery to a wastewater (sewage) treatment plant. Because industrial waste frequently goes down the same sewers as domestic and commercial nonindustrial waste, sewage often contains high levels of industrial chemicals and heavy metals like lead, mercury, cadmium, and arsenic.

Substances that are not removed by wastewater treatment processes are discharged via the treated effluent to a receiving stream, river, or coastal outlet. Inland waters ultimately reach the ocean, carrying with them some residual chemical that are not attenuated, stored, or degraded during their journey through the watershed. Other land-based sources of industrial pollutants in the ocean are pipeline discharges leaking underground storage tanks, and activities at ports and harbors. Intentional, illegal dumping in inland watersheds and in inland water bodies also can deliver industrial wastes to drainage ways, and ultimately to the ocean. In coastal watersheds, some industries discharge their wastes directly to the ocean. Industrial pollutants also can directly enter the ocean by accidental spills or intentional dumping at sea. Industrial chemicals can adversely affect the growth, reproduction, and development of many marine animals. Pollutants are appearing not only in the Pacific, Atlantic, and Indian Oceans and their marginal seas, but also in the more remote and once-pristine polar oceans. An array of contaminants have been found in the flesh of fish and marine mammals in polar regions. In addition to the environmental and ecological issues, there is growing concern over the potential human health impacts in aboriginal communities whose residents depend on fish and marine mammals for daily sustenance.

5.7 NOISE POLLUTION

Noise is unwanted sound and noise pollution can be defined as unwanted or offensive sounds that unreasonably intrude into our daily activities. It has many sources, most of which are associated with urban development: road, air and rail transport; industrial noise; neighborhood and recreational noise.

5.7.1 Sources of Noise

Road Traffic: Of all the sources of noise pollution, road traffic is the most prevalent and perhaps the most damaging source of noise pollution. More people are exposed to noise from motor vehicles than any other single source of noise and thus the impacts of traffic noise are a major factor in human society. The impact of road traffic noise on the community depends on various factors such as road location and design, land use planning measures, building design, vehicle standards and driver behavior.

Air Traffic : In the early days of aviation, the effects of noise were primarily considered in terms of their influence on an aircraft's passengers and crew. However, airline travel has since become much more common and widely available, resulting in an increased amount of air traffic and thus an increased number of people who are affected by aircraft noise, both has passengers and as members of communities located near airports. Noise from supersonic aircraft is of particular concern due to its intensity.

Railways : Railways like road traffic are a source of surface transportation noise. The level of noise associated with rail traffic is related to the type of engine or rolling stock used the speed of the train and track type and condition. The majority of noise emitted by trains is produced by the engines or by the interaction of wheels with the track. Other sources of noise in railways include warning signals at crossings, whistles and horns, freight classification yards, and railroad construction and maintenance equipment.

Industry : The four main categories of industrial activity cause noise pollution. They are product fabrication, product assembly, power generation and processing.

Noise is generated in all of these activities, with the majority occurring at the lower end of the frequency spectrum. While people around an industrial facility and the people within it are both affected by industrial noise, it is the workers within the plant the generally bear the brunt of most of it.

Construction: construction noise, a major source of noise pollution, is emitted by construction equipment. Like industrial equipment, construction equipment tends to produce more noise in the lower end of the frequency spectrum. But unlike industrial equipment, which emits noise that primarily affects workers within a facility, construction equipment tends to be used outdoors, and thus affects many other people other then the workers at the site.

Consumer Products : Consumer products represent a wide range of noise producing items categories such as

- Recreational (e.g. model airplanes, motorcycles, snowmobiles, go carts, all terrain vehicles, video arcades)
- Hobbies/workshop(e.g. chain saws, power saws, lawn mowers)
- Household (e.g. food blenders, vacuum cleaners, washers and dryers, air conditioners, and refrigerators)
- Music(e.g. personal stereos, rock concerts, and home stereos)

The least noisy consumer products belong to the household category, such as refrigerators and air conditioners, which don't exceed peak intensities of more than 75 dB.

Other Sources : In addition to the sources of noise pollution discussed thus far, there are several other sources, as well, such as sirens, agricultural noise, barking dogs, military noise, and noise generated by humans themselves. Since sirens are intended to convey argent messages, it is necessary for them to be loud enough to attract attention. Sirens are most often used by ambulances, police vehicles, and fire engines. Though obviously important for the safety of a community, sirens can also be considered bothersome, especially by those who live near hospitals, police stations, fire stations, or severe weather stations.

5.7.2 Effects of Noise

Subjected to 45 decibels of noise, the average person cannot sleep. At 120 decibels the ear registers pain, but hearing damage begins at a much lower level, about 85 decibels. The duration of the exposure is also important. Apart from hearing loss, noise can cause lack of sleep, irritability, heartburn, indigestion, ulcers, high blood pressure, and possibility heart disease. One burst of noise, as from a passing truck, is known to alter endocrine, neurological, and cardiovascular functions in many individuals; prolonged or frequent exposure to such noise tends to make the physiological disturbances chronic. In addition, noise induced stress creates severe tension in daily living and contributes to mental illness.

5.7.3 Industrlal Noise - Occupational Health Hazards

1. ***Noise-induced hearing loss:*** Exposure to sufficiently intense noise for a long enough duration results in damage to the inner ear and thus decreases one's ability to hear. In addition to a general decrease in the ability to detect sounds, the quality and clarity of auditory perception can be affected, as well. While these affects are often temporary, it is not uncommon for some residual permanent damage

to persist for the remainder of the effected person's life. Continued exposer to noise damages heasring The louder the noise, the less time it takes to cause hearing loss. There is fairly consistant evidence that prolonged exposure to noise levels at or above 85 dB(A) can cause deafness. The amount of deafness depends upon the degree of exposure. If we have do raise our voice while speaking to someone one meter away from us than our immediate environment could be hazardous to the health.

2. ***Non-auditory physiological effects:*** In addition to its effects on hearing, noise has also been implicated as having other physiological effects. Non auditory physiological effects of noise pollution include Cardiovascular and gastric effects, headaches, greater susceptibility to minor accidents, increased reliance on sedatives and sleeping pills and even increased mental hospital admission rates.

3. ***Effects on sleep:*** It is common that noise can disturb sleep. It has been found that in almost all cases increased levels of noise were associated with sleep disturbances; including shorter sleep diration, more frequent awakenings etc. Sleep disturbance by noise depends on the characteristics of the noise such as frequency, loudness and weather the noise is continuous or intermittent.

4. ***Annoyance:*** One of the most important effects of noise on humans is annoyance. A noise said to be annoying if an exposed individual or a group of individuals would reduce the noise, avoid, or leave the noisy area if possible.

5. ***Communication Interference:*** Noise pollution can have a considerable effect on communication. Noise can mask speech and even when speech is accurately understood, background noise may result in greater effort on the part of the speaker and listener than otherwise would be needed.

   Other effects: There are many other effects of noise other than those discussed so far, and some of these innumerable effects involve cognitive performance, occupational performance, psychological functioning and effects on aggression.

5.8 THERMAL POLLUTION

A large steam-electric power plant requires an enormous amount of cooling water. A typical nuclear plant, for example, warms about 40 m^3/s of cooling water by 10^0 C as it .passes through the plant's condenser. If that heat is released into a local river or lake, the resulting rise in temperature can dramatically affect life in the vicinity of the thermal plume. The major contributor to thermal pollution is the power industry. Both steam and nuclear powered generating stations use large amounts of water for cooling. Of the two, nuclear installations produce more thermal pollution as they are less thermally efficient, and cooling water is usually returned to its source after a temperature increase of 6-9^0 C.

As water temperature increases, two factors combine to make it more difficult for aquatic life to get sufficient oxygen to meet its needs. The first results from the fact that metabolic rates tend to increase with temperature, generally by about a factor of 2 for each 10°C rise in temperature, this causes an increase in the amount of oxygen required by organisms. At the same time, the available supplies of dissolved oxygen are reduced both because waste assimilation is quicker, drawing down DO at a faster rate, and because the amount of DO that the water can hold decreases with temperature. Thus, as temperatures increase, the demand for oxygen goes up while the amount of DO available goes down. The increase in water temperature may have deleterious effects on the local fauna and flora by decreasing the concentration of dissolved oxygen, which in turn increase their susceptibility to other pollutants.

5.8.1 Thermal Comforts

Human thermal comfort is defined as the state of mind that expresses satisfaction with the surrounding environment. Maintaining thermal comfort for occupants of buildings or other enclosures is one of the important goals of design engineers. Thermal comfort is affected by heat conduction, convection, radiation, and evaporative heat loss. Thermal comfort is maintained when the heat generated by human metabolism is allowed to dissipate, thus maintaining thermal equilibrium with the surroundings. Any heat gain or loss beyond this generates a sensation of discomfort. It has been long recognized that the sensation of feeling hot or cold is not just dependent on air temperature alone.

Thermal comfort is very important to many work-related factors. It can affect the distraction levels of the workers, and in turn affect their performance and productivity of their work.

The Factors determining thermal comfort include: personal factors (health, psychology, sociology & situational factors), air temperature, mean radiant temperature, air movement/velocity (see wind chill factor), relative humidity (see also perspiration), insulative clothing and activity levels.

5.8.2 Heat Island Effect

The term "heat island" describes built up areas that are hotter than nearby rural areas. The annual mean air temperature of a city with one million people or more can be warmer than its surroundings. Heat islands can affect communities by increasing summer time peak energy demand, air conditioning costs, air pollution and greenhouse gas emissions, heat-related illness and mortality, and water quality.

An urban heat island (UHI) is a metropolitan area which is significantly warmer than its surrounding rural areas. The phenomenon was first investigated and described by Luke Howard in the 1810s, although he was not the one to name the phenomenon. The temperature difference usually is larger at night than during the day, and is most apparent when winds are weak. Seasonally, UHI is seen during both summer and winter. The main cause of the urban heat island is modification of the land surface by urban development which uses materials which effectively retain heat. Waste heat generated by energy usage is a secondary contributor. As population centers grow they tend to modify a greater and greater area of land and have a corresponding increase in average temperature. The lesser-used term heat island refers to any area, populated or not, which is consistently hotter than the surrounding area.

Monthly rainfall is greater down wind of cities, partially due to the UHI. Increases in heat within urban centers increases the length of growing seasons, and decreases the occurrence of weak tornadoes. Increases in the death rate during heat waves has been shown to increase by latitude due to the urban heat island effect. The UHI decreases air quality by increasing the production of pollutants such as ozone, and decreases water quality as warmer waters flow into area streams, which stresses their ecosystems.

Not all cities have a distinct urban heat island. Mitigation of the urban heat island effect can be accomplished through the use of green roofs and the use of lighter-colored surfaces in urban areas, which reflect more sunlight and absorb less heat. Despite concerns raised about its possible contribution to global warming, any impact of the urban heat island on global warming is uncertain. Its impact on climate change has not been proved observationally or by any quantitative modeling, though recent qualitative speculations indicate that urban thermal plumes may contribute to variation in wind patterns that may itself influence the melting of arctic ice packs and thereby the cycle of ocean current.

5.8.3 Radiation Effects

- The losing of hair quickly and in clumps occurs with radiation exposure at 200 rems or higher.
- Since brain cells do not reproduce, they won't be damaged directly unless the exposure is 5,000 rems or greater. Like the heart, radiation

kills nerve cells and small blood vessels, and can cause seizures and immediate death.

- The certain body parts are more specifically affected by exposure to different types of radiation sources. The thyroid gland is susceptible to radioactive iodine. In sufficient amounts, radioactive iodine can destroy all or part of the thyroid. By taking potassium iodide, one can reduce the effects of exposure.
- When a person is exposed to around 100 rems, the blood's lymphocyte cell count will be reduced, leaving the victim more susceptible to infection. This is often referred to as mild radiation sickness.
- Intense exposure to radioactive material at 1,000 to 5,000 rems would do immediate damage to small blood vessels and probably cause heart failure and death directly.
- Radiation damage to the intestinal tract lining will cause nausea, bloody vomiting and diarrhea. This is occurs when the victim's exposure is 200 rems or more. The radiation will begin to destroy the cells in the body that divide rapidly. These including blood, GI tract, reproductive and hair cells, and harms their DNA and RNA of surviving cells.
- Because reproductive tract cells divide rapidly, these areas of the body can be damaged at rem levels as low as 200. Long-term, some radiation sickness victims will become sterile.

5.9 NUCLEAR POLLUTION

Any undesirable effect caused to the environment due to radioactive substances or radiations is called nuclear pollution. Major source is the Nuclear power plants. If traces of the radioactive substances are present in the water that is released from the plant, it will cause nuclear pollution. Emission of radiations can also cause this kind of pollution. It affects almost all life forms in the surrounding environment. From planktons to Human beings nothing is spared. To be more specific, the radiations can cause mutations that lead to cancer, and the dose of radiation or the level of pollution determines lethality or how deadly it is.

However, nuclear pollution is extremely hazardous in nature. It occurs as a result of nuclear explosions that are performed while conducting nuclear tests. These nuclear tests are carried out to invent better nuclear weapons. The explosions cause release of 15 to 20% radioactive material into the stratosphere. On entering this layer, they start falling into the earth's atmosphere. This fall can take any where from 6months to several years. 5% of these radioactive particles enter troposphere, which is the lowest layer of the atmosphere.

The smallest particles of the radioactive material are called fallout. The fallout settles on the leaves of plants and trees. These leaves are eaten by the grazing animals. Radioactive material now enters the ecosystem. Humans consume these particles through the process of food chain. Serious health problems now arise. Ingestion of radioactive material can lead to cancer and genetic mutation in humans. Fallouts that do not drop on leaves accumulate over the sea. This can be harmful for the sea life, which ultimately affects the humans.

It isn't necessary that only nuclear power stations cause nuclear pollution. Even other industries, not related to nuclear power production, can also contribute to it. Coal has small amounts of radioactive material in the form of uranium and thorium. These do not bum completely and become part of fly ash. Even while producing oil and gas, radium and similar elements are released in to the air.

Radioactive contamination or nuclear pollution is the most dangerous for the environment since the wastes maintain their radioactive properties for thousands of years. There is no way to have them assimilated in the soil, water or the air in the initial form. Reprocessing is solution we have to extent of nuclear pollution and clean the planet increasingly residues. The highest likelihood of radioactive elements reaching open environment is by accident during the transportation to the reprocessing plants located in some parts of the globe. Reprocessing in itself causes other pollution problems adding other risks to an already fragile environment condition.

Presently, no country has efficiently solved the issue of nuclear pollution in terms of radioactive waste storage. Every state would like to send the residues to some other place and be rid of them, while no truly viable conclusion is reached. Storage facilities as such require highly intransigent security and safety rules, periodical checks and regular updates on the storage environment. A responsible management of the nuclear waste would limit the risk of nuclear pollution on the long term, allowing us to live on a cleaner and safer planet, also preventing the temptation of dumping the waste in the oceans.

Nuclear pollution is not the only hazard that comes together with the use of radioactive energy: mass populations are jeopardized on a current basis if something happens to a reactor, as it was the case with the Russian Chernobyl for instance. There are other energy sources that are still highly effective without the huge risks of nuclear pollution or irradiation:

geothermal sources, ocean currents, tidal waves, wind and waterfalls, all make alternative power solutions that should not be neglected. Environment-friendly electricity is one of the chances this planet has to survive.

Fish and ocean plants are highly contaminated due to nuclear pollution; Greenpeace has repeatedly signaled out the huge amount of plutonium effluents produced by the nuclear plant on the coasts of England, for instance. Lobsters in the area have been found to be contaminated, hence the effects not only on humans but on the entire ecosystem is devastating. Attempts have been by an American company to even built a radioactive storage facility on Marshall Islands, ignoring the even higher potential threats for nuclear pollution under the circumstances of a growing sea level. Such solutions may appear convenient from a certain perspective, but when considered from a wider point of view, irresponsibility is obvious.

5.9.1 Nuclear Power Plants

Nuclear power is power, generally electrical produced from controlled, that is non-explosive nuclear reactions. Electric utility reactors heat water to produce steam, which is then used to generate electricity. In 2009, 15% of the world's electricity came from nuclear power, despite concerns about safety and radioactive waste management. More than 150 naval vessels using nuclear propulsion have been built. Nuclear fusion reactions are widely believed to be safer than fission and appear potentially viable, though technically quite difficult and have yet to be created on a scale that could be used in a functional power plant. Fusion power has been under intense theoretical and experimental investigation for many years. Both fission and fusion appear promising for some space propulsion applications in the mid- to distant-future, using low thrust for long durations to achieve high mission velocities. Radioactive decay has been used on a relatively small (few kW) scale, mostly to power space missions and experiments. As of 2005, nuclear power provided 2.1% of the world's energy and 15% of the world's electricity, with the U.S., France, and Japan together accounting for 56.5% of nuclear generated electricity.

5.9.2 Nuclear Radiation

Radiation is really nothing more than the emission of energy waves through space, as well as through physical objects. Usually these energy waves are electromagnetic radiation which are classified into Radio waves, Infrared waves, Visible light, Ultraviolet waves, X-ray, Gamma rays and Cosmic rays.

The actual radioactivity is a result of radioactive decay. The three types of radiation with varying abilities to penetrate objects or bodies are: Alpha, Beta, and Gamma radiation

You can shield yourself from alpha radiation by something as flimsy as a sheet of paper. Beta rays need six millimeters of aluminum and gamma rays are stopped by dense material only, like lead. These travel easily through an inch of lead. And the higher you are in the Earth's atmosphere the more exposed you are to these rays because the further they travel into our atmosphere the more they are slowed down. Astronauts are exposed to high levels of cosmic radiation.

5.9.3 Disasters and Impacts

It is considered to be the worst nuclear power plant disaster in history and the only level 7 event on the International Nuclear Event Scale. It resulted in a severe release of radioactivity following a massive power excursion that destroyed the reactor. Most fatalities from the accident were caused by radiation poisoning. On April 26, 1986 at 01:23 a.m. (UTC+3), reactor number four at the Chernobyl plant, near Pripyat in the Ukrainian Soviet Socialist Republic, had a fatal meltdown. Further explosions and the resulting fire sent a plume of highly radioactive fallout into the atmosphere and over an extensive geographical area, including the nearby town of Pripyat. Four hundred times more fallout was released than had been by the atomic bombing of Hiroshima. The plume drifted over large parts of the western Soviet Union, Eastern Europe, Western Europe, and Northern Europe. Rain contaminated with radioactive material fell as far away as Ireland. Large areas in Ukraine, Belarus, and Russia were badly contaminated, resulting in the evacuation and resettlement of over 336,000 people. According to official post-Soviet data, about 60% of the radioactive fallout landed in Belarus. The accident raised concerns about the safety of the Soviet nuclear power industry as well as nuclear power in general, slowing its expansion for a number of years while forcing the Soviet government to become less secretive. The countries of Russia, Ukraine, and Belarus have been burdened with the continuing and substantial decontamination and health care costs of the Chernobyl accident. It is difficult to accurately quantify the number of deaths caused by the events at Chernobyl, as over time it becomes harder to determine whether a death has been caused by exposure to radiation. The 2005 report prepared by the Chernobyl Forum, led by the International Atomic Energy Agency (IAEA) and World Health Organization (WHO), attributed 56 direct deaths

(47 accident workers and nine children with thyroid cancer) and estimated that there may be 4,000 (questioned, could be higher) extra cancer deaths among the approximately 600,000 most highly exposed people. Although the Chernobyl Exclusion Zone and certain limited areas remain off limits, the majority of affected areas are now considered safe for settlement and economic activity. The Chernobyl station is near the town of Pripyat, Ukraine, 18 km (11 mi) northwest of the city of Chernobyl, 16 km (10 mi) from the border of Ukraine and Belarus and about 110 km (68 mi) north of Kiev. The station consisted of four RBMK-1000 nuclear reactors, each capable of producing 1 gig watt (GW) of electric power, and the four together produced about 10% of Ukraine's electricity at the time of the accident. Construction of the plant began in the late 1970s, with reactor no. 1 commissioned in 1977, followed by no. 2 (1978), no. 3 (1981), and no. 4 (1983). Two more reactors, no. 5 and 6, also capable of producing 1 GW each, were under construction at the time of the disaster.

Nuclear power has developed into a mature and commercial reality with its part of success and threats. Although nuclear power can be used to generate electricity in an environment friendly manner, it is a subject of controversies owing to the threats related to it. Nuclear power has positive as well as negative impacts on people, society and environment Positive Impacts on People. The potential of nuclear power has been recognized by people. It is capable of providing unlimited and easily accessible energy. Due to the depletion in reserves of fossil fuels, costs of fuels and power is increasing at a fast pace which is directly affecting the people. Besides, the use of fossil fuels results in production of harmful oxides of carbon and nitrogen which have an adverse effect on the health of humans causing diseases like asthma, cancer, etc. Use of nuclear power does not produce these harmful gases which will thus reduce health problems of people. Also nuclear power is more economical than other sources like oil, coal, etc. Usage of nuclear power will help people to be independent in terms of energy and thus progress. France is comfortable using nuclear power for domestic as well as industrial purposes since it does not want to depend on other nations for its fuel supply and it is cheapest source of energy. The nuclear power can be used for electrification, industrial uses and heating purpose. Any organization can use nuclear power which is economic and more reliable than other energy sources. Also it is free of green house gases and thus helps an organization in meeting set standards regarding gas emissions with ease. Thus it results in enhanced productivity and economic growth.

5.9.4 Genetical Disorders

A **genetic disorder** is an illness caused by abnormalities in genes or chromosomes. While some diseases, such as cancer, are due in part to genetic disorders, they can also be caused by environmental factors. Most disorders are quite rare and affect one person in every several thousands or millions. Some types of recessive gene disorders confer an advantage in the heterozygous state in certain environments. A **single gene disorder** is the result of a single mutated gene. There are estimated to be over 4000 human diseases caused by single gene defects. Single gene disorders can be passed on to subsequent generations in several ways. Genomic imprinting and unapparent disomy, however, may affect inheritance patterns. The divisions between recessive and dominant types are not "hard and fast" although the divisions between autosomal and X-linked types are (since the latter types are distinguished purely based on the chromosomal location of the gene). For example, achondroplasia is typically considered a dominant disorder, but children with two genes for achondroplasia have a severe skeletal disorder that achondroplasics could be viewed as carriers of. Sickle-cell anemia is also considered a recessive condition, but heterozygous carriers have increased immunity to malaria in early childhood, which could be described as a related dominant condition.

5.10 SOLID WASTE

Solid waste generation, in one form or the other, is associated with human activities. Among the various problems resulting due to consequences of urbanization, management of solid waste is one. Management of municipal solid waste (MSW) in urban centers is becoming more complex due to scarcity of land for disposal (Babyrani devi and Bhoyar, 2003). Today, nearly half of the world's growing population lives in urban areas, causing enormous pressure on the local environment. Particularly, in the large agglomerations of the developing countries, inadequate waste management is the cause of serious urban pollution and health hazards. Affluent industrialized economies are facing an ever-increasing load of wastes and declining landfill space to dispose of these materials. Sustainable management of waste with the overall goal of minimizing its impact on the environment in an economically and socially acceptable way is a challenge for the coming decades (Christian Ludwig et al, 2003).

Large quantities of waste require proper systems of collection, transportation and disposal. It requires knowledge of what the wastes are made up of, how they need to be collected and disposed. Recycling of wastes and energy generation from waste also has great potential. Proper solid waste management requires commitment as well as resources to deal with wastes (B.K.Sharma and H.Kaur, 1996-1997). Solid waste management comprises of purposeful and systematic control of the generation, storage, collection, transport, separation, processing, recycling, recovery and disposal of the solid waste. An integrated solid waste management plan should include various components v.i.z., waste segregation and minimization, material recovery, waste processing and transformation, and its disposal on land. Solid waste management may thus be defined as the discipline associated with the control of generation, storage, collection, transfer and transport, processing and disposal of solid wastes in a manner that is in accord with the best principles of public health, economics, engineering, conservation, aesthetics and other environmental considerations, and that is also responsive to public attitudes (George Tchobanoglous, 1993).

5.10.1 Classification of Solid Waste

Solid waste is the material generated from various human activities and which is normally disposed as useless and unwanted. Hence, "Any unwanted or discarded material from residential, commercial, industrial, mining and agricultural activities that cause environmental problems may be termed as solid waste". It consists of the highly heterogeneous mass of discarded materials from the urban community as well as the more homogeneous accumulation of agricultural, industrial and mining wastes (C.S. Rao, 1991). Improper handling of society's byproducts in the form of domestic, commercial and industrial wastes can pose direct threats to both the public health and the quality of environmental resources.

In order to plan, design and operate a solid waste management system, a thorough knowledge of the quantities generated, the composition of the wastes and its characteristics are essential. There are many terms, which relate to the types and sources of wastes. A comprehensive classification of wastes is described below (Manual on Municipal Solid waste Management, 2000). The classification of solid wastes, their sources and description are given in Table 5.8.

Table 5.8 Classification of Solid waste.

Types of solid waste	Description	Sources
Food waste (garbage)	Wastes from the preparation, cooking & serving of food. market reuse, waste from the handling storage, & sale of meat & vegetables	Households, institutions, & commercials such as hotels, stores, restaurants, markets etc.
Rubbish	Combustible (primary organic), paper, cardboard, cartons, wood, boxes, plastics, rags, cloth, bedding, leather, rubber, grass, yard trimmings. Noncombustible(inorganic) metals, tin cans, metal foils, dirt, stones, bricks, ceramics, crockery, glass bottles, other mineral reuse.	
Ashes & residues	Residue from fires used for cooking & for heating buildings, cinders, clinkers, and thermal power plants.	
Bulky wastes	Large auto parts, tyres, stoves, refrigerators, other large appliances, furniture, large crates, trees, branches, palm fronts, stumps, floatage.	
Street waste	Street sweepings, dirt, leaves, catch basin dirt, animal droppings, contents of litter receptacles, dead animals.	Streets, side walks, alleys, vacant lots etc.
Dead animals	Small animals: cats, dogs, poultry, etc. large animals: horses, cows etc.	
Construction & demolition waste	Lumber, roofing & sheathing scraps, crop residues, broken concrete, rubble, plaster, conduit, pipe, wire, insulation etc.	Construction & demolision sites, remodeling, repairing sites
Industrial waste & sludges	Solid waste resulting from industry processes & manufacturing operations, such as food processing wastes, boiler house cinders, wood, plastic & metal scraps & shaving etc. effluent treatment plant sludge of industries & sewage treatment plant sludges, coarse screening, grit & septic tank.	Factories, power plants, treatment plants, etc.
Hazardous wastes	Hazardous wastes, pathological wastes, explosives, radioactive materials, toxic wastes etc.	Households, hospitals, institutions, stores, industry etc.

Source : Solid waste management in developing countries by Bhide and Sundaresan, INSDOC, New Delhi, 1987

Municipal Waste: Municipal waste includes waste resulting from municipal activities and services such as street wastes, dead animals, market wastes and abandoned vehicles. However, the term is commonly applied in a wider sense to incorporate domestic wastes and commercial wastes.

Domestic / Residential Waste: This category of waste comprises the solid wastes that originate from single and multi-family house hold units. These wastes are generated as a consequence of house hold activities such as cooking, cleaning, repairs, hobbies, redecoration, empty containers packaging, clothing, old books, paper and old furnishings.

Commercial Waste: Included in this category are solid wastes that originate in offices, wholesale and retail stores, restaurants, hotels, markets, warehouses and other commercial establishments. Some of these wastes are further classified as garbage and others as rubbish.

Garbage: Garbage is the term applied to animal and vegetable waste resulting from the handling, storage, sale, cooking and serving food. Such wastes contain putrescible organic matter, which produces strong odours and therefore attracts rats, flies and other vermin. It requires immediate attention in its storage, handling and disposal.

Rubbish: Rubbish is general term applied to solid wastes originating in households, commercial establishments and institutions, excluding garbage & ashes.

Institutional Waste: Institutional wastes are those arising from institutions such as schools, universities, hospitals and research institutes. It includes wastes, which are classified as garbage and rubbish, as well as wastes, which are considered to be hazardous to public health and to the environment.

Ashes: Ashes are the residues from the burning of wood, coal, charcoal, coke and other combustible materials for cooking and heating in houses, institutions and small industrial establishments. When produced in large quantities at power generation plants and factories, these wastes are classified as industrial wastes. Ashes consist of a fine powdery residue, cinders and clinker often mixed with small pieces of metal and glass.

Bulky Wastes: In this category are bulky household wastes, which can't be accommodated in the normal storage containers of households. For this reason they require special collection. In developed countries residential bulky wastes include household furniture and "white goods" appliances such as stoves, washing machines and refrigerators, mattresses and springs, rugs, TV sets, water heaters, tires, lawn mowers, auto parts, tree and brush debris, and so forth. Commercial bulky wastes include packaging and containers in a wide range of sizes, including corrugated cardboard, and wood boxes, fiber, plastic and steel drums usually under 40 gallons (0.15m^3), loose and bundled paper (office, printouts), bundles of textiles and plastics, bales of corrugated and paper, furniture and equipment, and flat and wire banding. Industrial bulky waste includes dunnage, including crates, cartons, pallets, skids; large and small steel, fiber, and plastic drums; bales and rolls of paper, plastics, and textiles; miscellaneous metal boxes, tubing, rod, punchings, and skeleton; wire, rope, and metal banding; and paper, textile, and plastic streamers (William D. Robinson, 1986).

Street Sweeping: This term applies to wastes that are collected from streets, walkways, alleys, parks and vacant lots. In the more affluent countries manual street sweeping has virtually disappeared but it still commonly takes place in developing countries, where littering of public places is a far more widespread and acute problem. Street wastes include paper, cardboard, plastic, dirt, dust, leaves and other vegetable matter.

Dead Animals: This is term applied to dead animals that die naturally or accidentally killed. This category does not include carcass and animal parts from slaughterhouses, which are regarded as industrial wastes. Dead animals are divided into 2 groups, large and small. Among the large animals are Horses, Cows, Goats, Sheep and the like. Small animals include dogs, cats, rabbits and rats. The reason for this differentiation is that large animals require special equipment for lifting and handling during their removal. If not collected promptly, dead animals are a threat to public health because they attract flies and other vermin as they putrify. Their presence in public places is particularly offensive and emits foul smell from the aesthetic point of view.

Construction and Demolition Wastes: Construction and demolition wastes are the waste materials generated by the construction, refurbishment, repair and demolition of houses, commercial buildings and other structures. It mainly consists of earth, stones, concrete, bricks, lumber, roofing materials, plumbing materials, heating systems and electrical wires and parts of general municipal waste stream, but when generated in large amounts at building and demolition sites, it is generally removed by contractors for filling low lying areas and by urban local bodies for disposal at landfills. While retrievable items such as bricks, wood metal are recycled, the concrete and masonry waste accounting for 50% of the waste from construction and demolition activities, are not been currently recycled in India. Concrete and masonry waste can be recycled by sorting, crushing and sieving into recycled aggregates. These recycled aggregates can be used to make concrete for road construction and building material.

This category waste is complex due to the different types of building materials being used but in general may comprise of major components like Cement concrete, Bricks, Cement plaster, Steel (from RCC, door/ window frames, roofing support etc., Rubble, Stone (marble, granite, sand stone), Timber/wood and a few minor components like Conduits (iron, plastic), Pipes (GI, iron, plastic), Electrical fixtures (copper/ aluminum wiring, wooden baton, bakelite, wire insulation, plastic switches), Panels (wooden, laminated), Others (Glazed tiles, glass panels).

Industrial Wastes

In this category are the discarded solid material of manufacturing processes and industrial operations. They cover a vast range of substances which are unique to each industry. For this reason they are considered separately from municipal wastes. However, solid wastes from small industrial plants and ash from power plants are frequently disposed of at municipal landfills. The major generators in the industrial solid wastes are the thermal power plants producing coal ash, the integrated Iron and steel mills producing blast furnace slag and steel melting slag, non-ferrous industries like aluminum, zinc, and copper producing red mud and tailings, sugar industries generating press mud, pulp and paper industries producing lime and fertilizer and allied industries producing gypsum. The source and quantum of generation of some major industrial wastes is given in Table 5.9.

Table 5.9 Source and Quantum of Generation of Some Major Industrial Waste.

S.No	Name	Quantity (million tones per annum)	Source/origin
1	Steel &blast furnace	35.0	Conversion of pig iron to steel& manufacture of iron
2	Brine mud	0.02	Caustic soda industry
3	Copper slag	0.0164	By product from smelting of copper
4	Fly ash	70.0	Coal based thermal power plants
5	Kiln dust	1.6	Cement plants
6	Lime sludge	3.0	Sugar, paper, fertilizer tanneries, soda ash, calcium carbide industries
7	Mica scraper waste	0.005	Mica mining areas
8	Phosphogypsum	4.5	Phosphoric acid plant, ammonium phosphate
9	Red mud/ bauxite	3.0	Mining & extraction of alumina from bauxite
10	Coal washery dust	3.0	Coal mines
11	Iron tailing	11.25	Iron ore
12	Lime stone wastes	50.0	Lime stone quarry

Source : National waste management council- Ministry of environment and forests 1990/1999

Hazardous Wastes: Hazardous wastes may be defined as wastes of industrial, institutional or consumer origin which because of their physical, chemical or biological characteristics are potentially dangerous to human and the environment. In some cases although the active agents may be liquid or gaseous, they are classified as solid waste because they are confined in solid containers. Typical examples are solvents, paints and pesticides whose spent containers are frequently mixed with municipal wastes and become part of urban waste stream. Table 5.10 lists typical hazardous products found in co-mingled with municipal solid waste. Certain hazardous wastes cause explosions in incinerators and fires at landfill sites. Others, such as pathological wastes from hospitals and radioactive wastes, require special handling at all time. Good management practice should ensure that hazardous wastes are stored, collected, transported and disposed off separately, preferably after suitable treatment to render them innocuous.

Table 5.10 Hazardous products found in co-mingled with municipal solid waste.

Product	Concern
Household cleaners: Chlorine bleach, Furniture polish, Glass cleaners, Outdated medicines, Shoe polish, Silver polish, Spot remover, Drain openers etc.	Corrosive, Flammable and Irritants
Personal Care products: Hair waving lotions, Medicated shampoos, Nail polish remover, Rubbing alcohol	Poisonous and Flammable
Automotive Products: Antifreeze, Brake and transmission fluid, Car batteries, Diesel fuel, Kerosene, Gasoline and Waste oil	Flammable, Poison and Corrosive
Paint Products: Enamel, Paints and paint solvents and thinners	Flammable
Miscellaneous Products: Batteries, Photographic chemicals, House plant insecticides, herbicides and chemical fertilizers	Corrosive and Poisonous

Source: Tchobanoglous, G., Theisen, H., and Vigil, S.A. (1993), "Integrated Solid Waste Management – Engineering Principles and Management Issues" Mc. Graw Hill, New York.

Sewage Wastes: The solid by-products of sewage treatment are classified as sewage wastes. They are mostly organic and derive from the treatment of organic sludge from both the raw and treated sewage. The inorganic fraction of the raw sewage such as grit is separated at a preliminary stage of treatment, but because it entrains putrescible organic mater which may contain pathogens, must be buried / disposed off without delay. The bulk of treated dewatered sludge is useful as a soil conditioner but invariably its use for this purpose is uneconomical. The solid sludge therefore enters the stream of municipal wastes unless special arrangements are made for its disposal.

Biomedical / Hospital Waste: Hospital waste is generated during the diagnosis, treatment, or immunization of human beings or animals or in research activities in these fields or in the production or testing of biologicals. It may include wastes like sharps, soiled waste, disposables, anatomical waste, cultures, discarded medicines, chemical wastes, etc. These are in the form of disposable syringes, swabs, bandages, body fluids, human excreta, etc. This waste is highly infectious and can be a serious threat to human health if not managed in a scientific and discriminate manner. It has been roughly estimated that of the 4 kg of waste generated in a hospital at least 1 kg would be infected.

These wastes are categorized into 10 different categories as:

- Human anatomical waste (tissues, organs, body parts etc.)
- Animal waste
- Microbiology and biotechnology waste, such as, laboratory cultures, microorganisms, human and animal cell cultures, toxins etc.
- Waste sharps such as, hypodermic needles, syringes, scalpels, broken glass etc
- Discarded medicines and cyto-toxic drugs
- Soiled waste, such as dressings, bandages, plaster casts, material contaminated with blood etc
- Solid waste (disposal items like tubes, catheters etc., excluding sharps)
- Liquid waste generated from any of the infected areas
- Incineration ash
- Chemical waste

Surveys carried out by various agencies show that the health care establishments in India are not giving due attention to their waste management. After the notification of the Bio-medical Waste (Handling and Management) Rules, 1998, these establishments are slowly streamlining the process of waste segregation, collection, treatment, and disposal. Many of the larger hospitals have either installed the treatment facilities or are in the process of doing so.

Plastics: Plastics, due to their versatility in use and impact on environment can be grouped under a different category of solid waste. Plastic with its exclusive qualities of being light yet strong and economical, has invaded every aspect of our day-to-day life. It has many advantages viz., durable, light, easy to mould, and can be adapted to different user requirements. Once hailed as a 'wonder material', plastic is now a serious worldwide environmental and health concern, essentially due to its non-biodegradable nature.

In India, the plastic industry is growing phenomenally. Plastics have use in all sectors of the economy – infrastructure, construction, agriculture, consumer goods, telecommunications, and packaging. The sources of generation of waste plastics are given in Table 5.11. Along with a growth in the use, a countrywide network for collection of plastic waste through rag pickers, waste collectors and waste dealers and recycling enterprises has sprung all over our country over the last decade. More than 50% of the plastic waste generated in our country is recycled and used in the manufacture of various plastic products. Impacts of plastics on the environment are extremely wide ranging. These include,

- Careless disposal of plastic bags chokes drains
- It blocks the porosity of the soil
- It causes problems for groundwater recharge
- Plastic disturbs the soil microbe activity
- Once ingested, can kill animals
- They contaminate foodstuffs due to leaching of toxic dyes and transfer of pathogens.

Table 5.11 Source of generation of waste plastics.

Household	Carry bags Bottles Containers Trash bags
Health and Medicare	Disposable syringes Glucose bottles Blood and uro bags Intravenous tubes Catheters Surgical gloves
Hotel and Catering	Packaging items Mineral water bottles Plastic plates, glasses, spoons
Air/Rail Travel	Mineral water bottles Plastic plates, glasses, spoons Plastic bags

Source : http://edugreen.teri.res.in/explore/solwaste/plastics.htm

In fact, a major portion of the plastics i.e. approximately 60-80% of the plastic waste generated in India is collected and segregated to be recycled. The rest remains strewn on the ground, littered around in open drains, or in unmanaged garbage dumps. Though only a small percentage lies strewn, it is this portion that is of concern as it causes extensive damage to the environment.

5.10.2 Factors Affecting the Solid Waste Generation

The factors which affects the quality and quantity of solid waste generated includes,

- Geographic location
- Seasonal factors
- Collection frequency
- Use of waste grinders
- Characteristics of population (standard of living, their cultural habits)
- Extent of salvaging and recycling
- Public attitudes
- Legislation

5.10.3 Composition and Characteristics of Solid Waste

The composition and characteristics of municipal solid wastes vary through out the world. Even in the same country it changes from place to place as it depends on number of factors such as social customs, standard of living, geographical location, climate etc. MSW is heterogeneous in nature and consists of a number of different materials derived from various types of activities.

For example, data for different degrees of national wealth (annual per-capita income) are presented in Table 5.12. Waste composition also varies with socio-economic status within a particular community, since income determines life-style - consumption and cultural behavior.

Table 5.12 Patterns of Composition, Characteristics and Quantities.

Composition (% by weight)	Low income countries(1)	Middle income countries (2)	High income countries (3)
Metal	0.2-2.5	1-5	3-13
Glass, ceramics	.5-3.5	1-10	4-10
Food & garden waste	40-65	20-60	20-50
Paper	1-10	15-40	15-40
Textiles	1-5	2-10	2-10
Plastics/rubber	1-5	2-6	2-10
Misc. combustible	1-8	---	---
Misc. incombustible	---	---	---
Inert	20-50	1-30	1-20
Density (Kg/cubic meter)	250-500	170-330	100-170
Moisture content (% by weight)	40-80	40-60	20-30
Waste generation (KG/CAP/DAY)	0.4-0.6	0.5-0.9	0.7-1.8

Countries having percapita income less than US $360 (1978 prices)

Countries having a percapita income US $3500 (1978 prices)

Countries having percapita income US $360-3500 (1978 prices)

Source: Holmes, j: Managing solid waste in developing countries.

5.10.4 Collection of Solid Waste

Collection is the first fundamental function of solid waste management. Solid waste collection refers to the gathering of solid waste from place to place such as residential, commercial, industrial and institutional areas as well as public parks.

5.10.4.1 Methods of Collection

There are generally two methods of collection,

(i) ***Hauled container system :*** In this system the container is hauled from the collection point to the final point of disposal, processing facility or transfer station.

(ii) ***Stationary container system :*** In this system the container is emptied into the collection vehicles at the point of collection.

5.10.4.2 Layout of Collection Route

To manage a solid waste collection system effectively, the community, industrial park and the like should be divided into collection districts, in which a number of collection routes must be laid out. Layout of routes should follow a minimum of guidelines, among which could be the following.

- Route should not overlap and should be continuous.
- On the job time of the collection routes should be made equal as possible.
- Congested areas should not be collected during rush hours.
- To service a dead end street, collect it when it is on the right of the main street, to avoid making a left turn.
- As practicable as possible, collect a street on both sides.
- As practicable as possible, select a consistent collection pattern.
- For collection on steep streets, it is best for the vehicle to be collecting while it is moving down slope rather than up slope. This is safer and will save energy and time.

5.10.4.3 Door Step Collection

Under the principle of users pay, beneficiaries pay, and polluters pay, MCH has introduced the scheme of collecting user charges from bulk garbage generators in the city. 1700 establishments like hotels, restaurants, function halls, hospital, markets and commercial complexes etc., generating bulk garbage are identified and classified into 12 categories for levying user charges. During the year 2003-2004, 20 transport work packages were proposed to be launched for providing doorstep collection arrangement to these establishments.

5.10.5 Transportation of Solid Waste

As the distance from the collection system to the processing facility or disposal site increases, the cost of hauling or transportation also increases. There will be eventually a certain transport distance, where management must decide whether or not a transfer station is to be built.

There are two types of transportation of waste:

(i) ***Direct discharge transfer :*** In this, collection vehicles dump their loads directly into the larger transportation vehicles.

(ii) **Storage transfer station:** In this, the solid wastes are emptied into storage pits or platforms. The wastes are then later loaded into big transport vehicles for hauling to the destination point.

5.10.5.1 Transfer Stations

The transfer station is a transit point in the movement of garbage to the landfill site or solid waste processing plant. Here the garbage is brought in small dumper placer bins and tippers. The contents are transferred directly into large big tipper vehicles through a specially designed hopper. These big tipper trucks in turn take the garbage to the landfill site or MSW processing plant for final disposal.

The advantages of having a transfer station are,

- The small dumper placer vehicles need not have to travel long distances up to the landfill site which on an average are located at 20 to 25Km away from the center of the city. This saves travel time and the fleet can be better utilized for making extra trips resulting in effective cleaning and sweeping.
- There is saving on the consumption of the fuel and as such the cost of the garbage transport is minimized.
- The wear and tear of the tyres and other components of vehicles are minimized by avoiding long trips and adverse conditions at landfill sites.
- There will be less traffic at landfill site thereby facilitating proper spreading of garbage and giving a better look at these sites.

5.10.5.2 Segregation of Solid Waste

Collection service structure providing community with waste bins conveniently placed for the people to deposit domestic waste, and door to door collection of waste along with separation of waste at source, into biodegradable and non-biodegradable components would not only reduce the cost of transportation for final disposal but also provide segregated organic waste stock for energy activities. Separating waste is essential as the amount of waste being generated today causes immense problems. The objectives of sorting are,

- To separately store recyclable material for reuse
- To ensure that waste which can be processed for recovery of material and energy does not become co-mixed with undesirable elements
- To separately store hazardous waste for disposal in hazardous waste landfills or for appropriate processing
- To minimize the waste and ensure reduction in landfill space for final disposal.

Waste can be segregated into the following streams,

1. ***Biodegradable Wastes:*** Biodegradable waste includes organic waste, e.g. kitchen waste, vegetables, fruits, flowers, leaves from the garden, and paper.
2. ***Non-biodegradable Wastes:*** Non-biodegradable waste can be further segregated into:
   (a) Recyclable waste – plastics, paper, glass, metal, etc.
   (b) Toxic waste – old medicines, paints, chemicals, bulbs, spray cans, fertilizer and pesticide containers, batteries, shoe polish.
   (c) Soiled – hospital waste such as cloth soiled with blood and other body fluids. Toxic and soiled waste must be disposed of with utmost care.

Certain items are not biodegradable but can be reused or recycled. In fact, it is believed that a larger portion can be recycled, a part of it can be converted to compost, and only a smaller portion of it is real waste that has no use and has to be discarded. According to the Municipal Solid Wastes (Management and Handling) Rules, 2000, wastes must be segregated into three categories as given in Table 5.13.

Table 5.13 Segregation of Waste.

Organic Wastes	Recyclable Wastes	Others
Tea leaves Egg shells Old food Vegetable peels Meat Bones	Paper Shampoo bottles Glass Note books Wires Safety pins Caps of mineral water bottles	PET mineral water bottles Nitrogen sealed packaging for chips Terapacks Thermocol Carbon paper Plastic coated visiting cards Sachets

Source : A Support Manual for Municipal Solid Wastes (Management and Handling) Rules, 2000, CPCB, MOEF, Delhi.

5.10.5.3 Methods of Segregation

Segregation or sorting can be carried out manually or through semi-mechanised and fully mechanized systems.

Manual sorting operation: *Manual sorting operation comprises of,*

- Unloading the waste
- Manually (with protective measures) spreading the waste
- Hand picking (with protective measures) visually identifiable waste for reuse
- Collecting and stockpiling the remaining waste

Semi-mechanized sorting operation: *Semi-mechanized sorting operations consists of,*

- Unloading of waste (mechanized)
- Loading of waste on conveyor belts (mechanized)
- Hand picking of visually identifiable waste off the belt for reuse (manual)
- Collecting, stocking and reloading the remaining waste (mechanized)

Fully-mechanized sorting operation: *Fully-mechanized sorting operations comprise of,*

- Unloading of waste
- Size reduction of waste through shredders and crushers
- Size separation of waste using screening devices
- Density separation of waste
- Magnetic separation of waste
- Compaction of waste through balers/crushers
- Reloading of waste

5.10.6 e-waste

Electronic waste, popularly known as 'e-waste' can be defined as electronic equipments / products connects with power plug, batteries which have become obsolete due to: advancement in technology changes in fashion, style and status nearing the end of their useful life.

E-waste encompasses ever growing range of obsolete electronic devices such as computers, servers, main frames, monitors, TVs & display devices, telecommunication devices such as cellular phones & pagers, calculators, audio and video devices, printers, scanners, copiers and fax machines besides refrigerators, air conditioners, washing machines, and microwave ovens, e-waste also covers recording devices such as DVDs, CDs, floppies, tapes, printing cartridges, military electronic waste, automobile catalytic converters, electronic components such as chips, processors, mother boards, printed circuit boards, industrial electronics such as sensors, alarms, sirens, security devices, automobile electronic devices.

Electronic waste or e-waste is one of the rapidly growing environmental problems of the world. In India, the electronic waste management assumes greater significance not only due to the generation of our own waste but also dumping of e-waste particularly computer waste from the developed countries. With extensively using computers and electronic equipments and people dumping old electronic goods for new ones, the amount of E-Waste generated has been steadily increasing. At present Bangalore alone generates about 8000 tonnes of computer waste annually and in the absence of proper disposal, they find their way to scrap dealers. E-Parisaraa, an eco-friendly recycling unit on the outskirts of Bangalore which is located in Dobaspet industrial area, about 45 Km north of Bangalore, makes full use of E-Waste. The plant which is India's first scientific c-waste recycling unit will reduce pollution, landfill waste and recover valuable metals, plastics & glass from waste in an eco-friendly manner. E-Parisaraa has developed a circuit to extend the life of tube lights. The circuit helps to extend the life of fluorescent tubes by more than 2000 hours. If the circuits are used, tube lights can work on lower voltages. The initiative is to aim at reducing the accumulation of used and discarded electronic and electrical equipments. India as a developing country needs simpler, low cost technology keeping in view of maximum resource recovery in an environmental friendly methodologies. E-Parisaraa, deals with practical aspect of e-waste processing as mentioned below by hand. Phosphor affects the display resolution and luminance of the images that is seen in the monitor.

REVIEW QUESTIONS

1. Write notes on classification and sources of pollution.
2. Write notes on causes and effects of pollution.
3. List out all industries causing different types of pollution.
4. Explain primary and secondary air pollutants.
5. Write short notes on (i) Automobile pollution (ii) Industrial pollution (iii) Ambient air quality standards.
6. What are the causes of water pollution? Explain.
7. Explain major pollutants of water and their sources.
8. Discuss in detail the water quality requirements for different uses.
9. Explain in detail the waste water treatment methods.
10. Write notes on (i) Sewage Treatment Plant (STP) (ii) CETP
11. Explain Soil pollution.
12. Discuss the issues related to Marine pollution.
13. Explain sources of noise pollution and their standards of control.
14. Write notes on Thermal comforts and factors of determining thermal comforts.
15. Explain Heat Island effect, Radiation effects.

16. Explain Nuclear pollution due to nuclear power plants.
17. Write about types of wastes.
18. What are the factors affecting the solid waste generation rates?
19. Describe composition and characteristics of solid waste.
20. Explain various methods of collections and management of solid waste.
21. What do you understand e-waste? Discuss.

CHAPTER 6

GLOBAL ENVIRONMENTAL PROBLEMS AND GLOBAL EFFORTS

6.1 GREEN HOUSE EFFECT

The "greenhouse effect" is the heating of the Earth due to the presence of greenhouse gases. It is named this way because of a similar effect produced by the glass panes of a greenhouse. Shorter-wavelength solar radiation from the sun passes through Earth's atmosphere, then is absorbed by the surface of the Earth, causing it to warm. Part of the absorbed energy is then reradiated back to the atmosphere as long wave infrared radiation. Little of this long wave radiation escapes back into space; the radiation cannot pass through the greenhouse gases in the atmosphere. The greenhouse gases selectively transmit the infrared waves, trapping some and allowing some to pass through into space. The greenhouse gases absorb these waves and reemits the waves downward, causing the lower atmosphere to warm (Fig. 6.1)

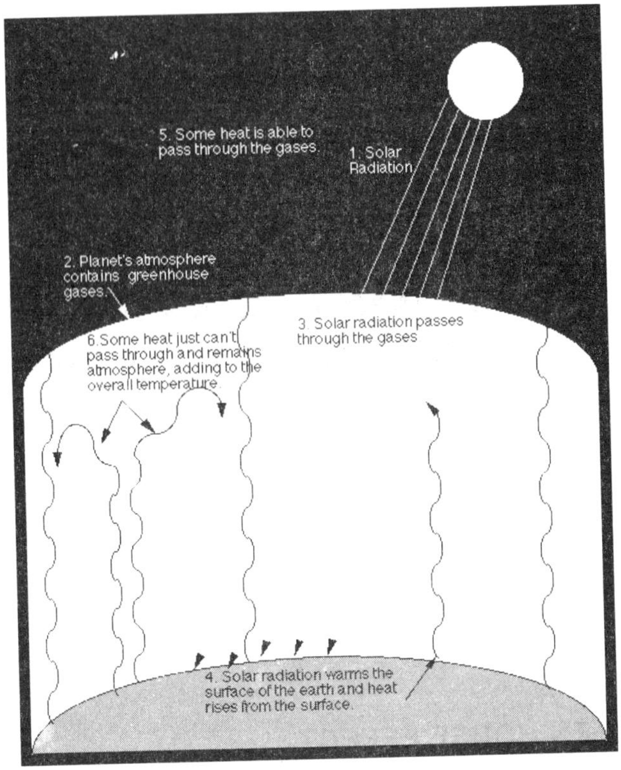

Fig. 6.1 Process of global warming and how greenhouse gases create the "greenhouse effect"

6.2 GREENHOUSE GASES

The three most powerful long lived greenhouse gases in the atmosphere are carbon dioxide, methane, and nitrous oxide. Water vapor has approximately twice the effect of the second most powerful greenhouse gas-carbon dioxide. Human activities do not have any significant direct impact on the level of water vapor in the atmosphere. However, as a result of global warming it is likely that human activities will have a significant indirect impact on the level of water vapor in the atmosphere. Water vapor is the most important greenhouse gas and the development of a better understanding of the effect of global warming on atmospheric water in all its forms (solid, liquid, and gas) is of critical importance.

Fig. 6.2 shows the concentrations of carbon dioxide, methane, and nitrous oxide in the atmosphere over the past 1000 years as measured in air trapped in bubbles in ice core samples from sites in Antarctica and Greenland and, for the more recent data, measured directly in ambient air.

The units for carbon dioxide concentration are parts per million (ppm) while those for methane and nitrous oxide are parts per billion (ppb), 1 ppm = 1000 ppb. The current atmospheric concentration of carbon dioxide is 370 ppm, meaning that in samples of one million molecules of ambient air, on average, 370 molecules of carbon dioxide can exist. As seen from Fig 6.2 the overall trend of carbon dioxide, methane, and nitrous oxide concentrations is the same. For the time period 1000–1800 there is little, or no, discernable change, while over the period 1800–2000 there is a substantial increase in the atmospheric levels of all three gases. It is well established that the increases in the atmospheric concentration of carbon dioxide, methane, and nitrous oxide during 1800–2000 reflect the impact of human activities in general and the industrial revolution in particular.

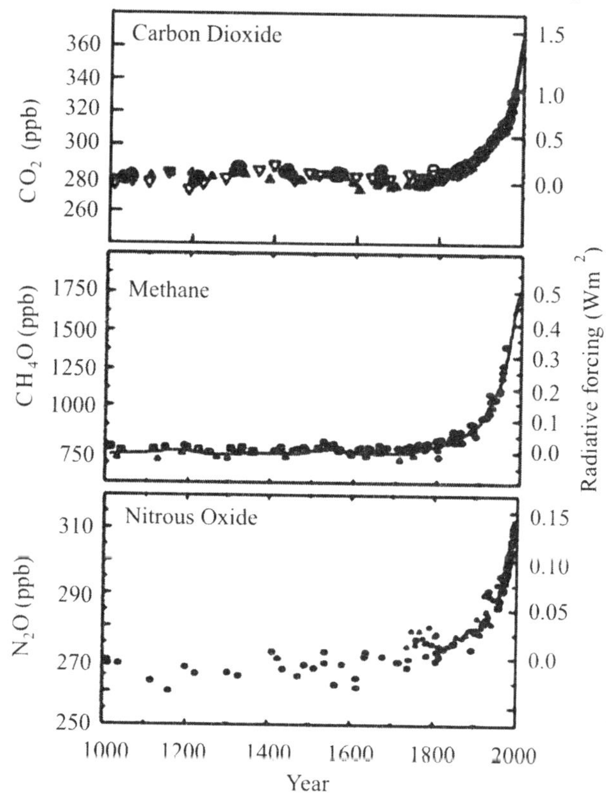

Fig 6.2 *Atmospheric concentration of carbon dioxide (CO_2), methane (CH_4), and nitrous oxide (N_2O) over the past 1000 years. Different symbols are measurements using ice cores from different sites in Antarctica and Greenland. The straight line in the CO_2 plot are the recent in-situ atmospheric measurements. The curve through the CH_4 data is a smoothed fit. (Reproduced with permission from IPCC)*

6.3 GLOBAL WARMING

Earth has become warmer over the last century. The Inter-governmental Panel "on Climate Change (IPCC), a group established by the World Meteorological Organization (WMO) and the United Nations Environment Programme (UNEP), reports that the average surface temperature of the earth has increased during the twentieth century by about 0.6°C ± 0.2°C (The ± 0.2°C means that the increase might be as small as 0.4° C or as great as 0.8°C) This may seem like a small shift, but although regional and short-term temperatures do fluctuate over a wide range, global temperatures are generally quite stable. In fact, the difference between today's average global temperature and the average global temperature during the last Ice Age is only about 5° C. Indeed, it is warmer today around the world than at any time during the past 1000 years, and the warmest years of the previous century have occurred within the past decade.

Human activities primarily the burning of fossil fuels have increased the greenhouse gas content of the earth's atmosphere significantly over the same period. Carbon dioxide is one of the most important greenhouse gases, which traps heat near the planet's surface.Measurements from a variety of sources have suggested that the earth's average atmospheric temperature has risen over the last several hundred years and there is no doubt that increases in the atmospheric concentration of carbon dioxide and other greenhouse gases strengthens the greenhouse effect and contributes to global warming or the climate change.

6.3.1 Causes of Global Warming

- Rapid use of fossil fuel
- Rapid use of fossil fuel will emit large amount of greenhouse gases, especially carbon dioxide. Deforestation/clearing of lands. The increase of human population coincidentally demands for more lands to use. This leads to clearance of forest area in many regions. By eliminating forests, carbon dioxide that actually should be photosynthesized is left in the atmosphere and accumulates to contribute to the increase. The use of CFC in (CFC electronic appliances decreases ozone (O_3) in the atmosphere, causing ozone depletion that increase the amount of solar radiation arriving to the earth.
- Open burning of trash. Open burning of trash worldwide emits greenhouse gases. Vehicle and industrial emissions, volcanic eruptions.

6.3.2 Signs of Global Warming

The receding of ice formations on earth (snows at mountain-tops, glaciers, and Antarctic and Arctic ice).The increase of shrubbery in Arctic. Thinner clouds over the sky that decrease the ability to reflect heat from the sun. The discovery of the decrease of earth's albedo (the amount of sunlight reflection by the earth surface to the moon) by 2.5 percent, which means the earth has lost some levels of capability to reflect sunlight to the moon. Change in wind directions

6.3.3 Impacts

Stormy weather (more chances for hurricanes, floods, cyclones, and storms to happen). Increased severity of drought, hunger and spread of diseases, especially in poor countries. Declines of amphibians, caused by altered precipitation patterns Resulted in lower levels of pond and lake waters, where

- Amphibians survive
- Damages to coral reefs
- Marine diseases
- Rising ocean temperature
- Ecosystem degradation
- Declining of biodiversity
- Economic and social downturn

6.4 SEA LEVEL RISE

Green house effect or global warming may cause a Sea Level Rise, which will have a great impact on the long-term coastal morphology. The possible and gradual Sea Level Rise will cause a general shoreline retreat and an increased flooding risk and has to be handled according to the local conditions. Sea level rise (SLR) due to climate change is a serious global threat.

6.5 CLIMATE CHANGE AND THEIR IMPACTS ON HUMAN ENVIRONMENT

The average temperature in many regions has been increasing in recent decades. The global average surface temperature has increased by 0.6^0 C - 0.20^0 C over the last century. Globally, 1998 was the warmest year and the 1990s the warmest decade on record. Many countries have experienced increases in rainfall, particularly in the countries situated in the mid-to high-latitudes. In some regions, such as parts of Asia and Africa, the frequency and intensity of droughts have been observed to increase in recent decades. Episodes of El Nino, which creates great storms, have been more frequent, persistent and intense since the mid-1970s compared with the previous 100 years. All these are signs that the Earth is ailing. Its climate is changing, making it more difficult for mankind to survive. The earth is losing its equilibrium due to the imbalances created by human activities.

Projections of future climate change are derived from a series of experiments made by computer based global climate models. These are calculated based on factors like future population growth ad energy use. Climatologists of the intergovernmental panel on climate change (IPCC) have reviewed the results of several experiments in order to estimate changes in climate in the course of this century. These studies have shown that in the near future the global mean surface temperature will rise by 1.4°C - 5.8^0 C. this 'warming' will be greatest over land areas, and at high latitudes. The projected rate extreme is likely to increase, leading to floods or drought. There will be fewer cold spells but more heat waves. The frequency and intensity of the El Nino is likely to increase. The global mean sea level is projected to rise by 9.88 cm by the year 2100. More than half of the world's population now lives within 60 km of the sea. They are likely to be seriously impacted by the ingress of salt water and by the rising sea. Some of the most vulnerable regions are the Nile delta in Egypt, the Ganges- Brahmaputra delta in Bangladesh, and many small islands including the Marshall islands and the Maldives, (WHO, 2001).

Human societies will be seriously affected by extremes of climate such as droughts and floods. A changing climate would bring about changes in the frequency and/or intensity of these extremes. This is also a fundamental concern for human health. To a large extent, public health depends on safe drinking water, sufficient food, secure shelter, and good social conditions. All these factors are affected by climate change. Freshwater supplies may be seriously affected, reducing the availability of clean water for drinking and washing during drought as well as floods. Water can be contaminated and sewage systems may be damaged. The risk of spread of infectious diseases such as diarrheal also indirectly through an increase in pests and plant or animal diseases. The local reduction in food production would lead to starvation and malnutrition with land-term health consequences, especially for children. Food and water shortages may lead to conflicts in vulnerable regions. Climate change related impacts on human health could lead to displacement of a large number of people, creating environmental refugees and lead to further health issues.

Changes in climate may affect the distribution of vector species (e.g., mosquitoes) which, in turn, will increase the spread of disease, such as material and filariasis, to new areas which lack a strong public health infrastructure. The seasonal transmission and distribution of many diseases that are transmitted by mosquitoes (dengue, yellow fever) and by ticks (Lyme disease, tick-borne encephalitis) may spread due to climate change.

6.6 OZONE DEPLETION AND OZONE DEPLETING SUBSTANCES (ODS)

6.6.1 Ozone Depletion

The stratospheric ozone layer protects Earth from harmful ultraviolet radiation. The depletion of this layer is another major atmospheric challenge that is due to human technology. Environmental scientists have traced the problem to a widely used group of chemicals: the CFCs. The ozone layer reduces the amount of UV-B reaching the earth's surface; and thining of this layer increase the U.V. - B light penetration. Increases in UV-B can be related to adverse effects on human health, terrestrial and aquatic plants, and a degradation of materials.

The biological response to increased ultraviolet radiation is often represented action spectrum. The relative ability of various wavelengths to cause biological harm can be understood from the action spectrum of DNA molecule (Fig.5.3).DNA is highly sensitive to shorter wavelengths of UV, and this part of the light is effectively shielded by ozone. For longer wavelengths in the UV-B portion of the spectrum, the ozone shield is less effective, but DNA is correspondingly less sensitive. However, the intermediate range of wavelengths, (300 to 315 nm) which are less shielded by ozone cause greatest damage to DNA.

UV-induced changes in DNA, are found to be responsible for skin cancer which may be either the more common nonmelanoma (basal cell or squamous cell carcinomaos), or the much more life threatening malignant melanoma (Fig.6.4).Non melanoma skin cancer is more prevalent among people with fair complexion, is most likely to occur on areas of the skin habitually receiving greatest exposure to the sun, and is associated with areas of the world having the greatest amount of sunshine.Though it is rarely fatal, it often causes disfigurement. In the United States, there are about 400 000 new cases of non melanoma each year, with a fatality rate of about 1 percent.

It was observed that at higher latitudes there is less surface level UV-B, and non melanoma rates decrease by roughly a factor of 2 for each 10^0 increase in latitude. Based on these observations it, was predicted that for every 1-percent decrease in ozone, the added UV-B exposure would cause the incidence rate of non melanoma to increase by approximately 2-3 percent. By simply moving 100 miles closer to the equator, UV-B exposure is expected to increase by an amount comparable to what would be expected by a 5% thinning of the Ozone layer.

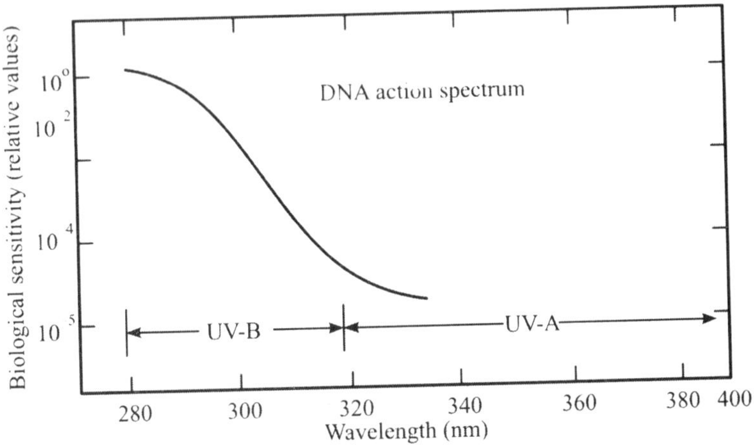

Fig. 6.3 Action spectrum for damage to the DNA molecule. The vertical scale is a relative Measure of biological sensitivity to radiation at each wavelength.

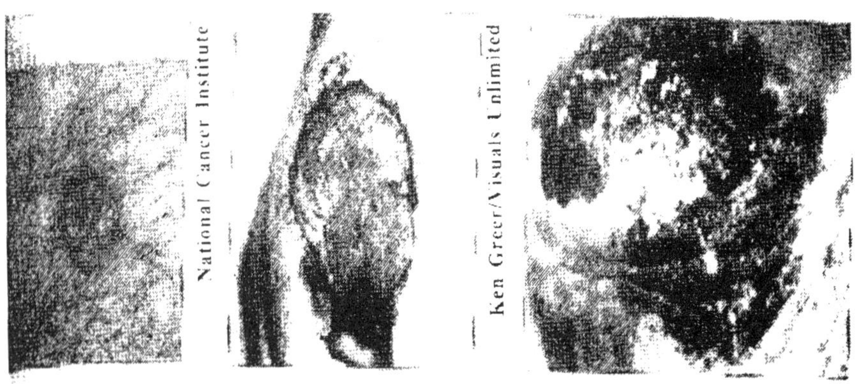

Fig. 6.4 Consequences of Ozone Depletion

Table 6.1 Medical conditions known to be induced by enhanced UV-B and their likelihood of increase with further stratospheric O3 depletion

Area affected	Condition or agent	Individual types	Radiation Amplification Factor*
Skin	Erythema sunburn		1.7
	Non-melanomatous cancer	Basal cell carcinoma Squamous cell carcinoma Lip cancer Salivary gland cancer	1.4
	Melanoma (cutaneous malignant)	Superficial spreading Lentigo maligna (Hutchinson's freckle) Nodular Unclassified	1.6
	Xeroderma pigmentosa	(conditioned worsened)	
	Skin ageing	E1astosis	1.2
Eye	Corneal damage	Acute Photokeratitis (snow-blindness)	1.1
	Lens Abnormalities	Nuclear cataracts Posterior subcapsular Cataracts Cortical cataracts Presbyopia (reading glasses required earlier)	0.7
	Other effects	Intraocular melanoma	
Infections	Viruses	Measles Chicken pox Herpes simplex HIV-I activation	
	Protozoa	Leishmaniasis Malaria	0.8
	Bacteria	Tuberculosis Leprosy	
	Fungi	Candiddiasis	

*Raf given 1991 UNEREPORT

6.6.2 Ozone Depleting Substances (ODS)

Ozone depleting substances (ODS) are those substances which deplete the ozone layer and are widely used in refrigerators, air-conditioners, fire extinguishers, in dry cleaning, as solvents for cleaning, electronic equipment and as agricultural fumigants. Ozone depleting substances include: Chlorofluorocarbons (CFCs), Halon, Methyl bromide, Carbon tetrachloride and Methyl chloroform

- **Chlorofluorocarbons (CFCs),** along with other chlorine- and bromine-containing compounds, have been implicated in the accelerated depletion of ozone in the Earth's stratosphere. CFCs were developed in the early 1930s and are used in a variety of industrial, commercial, and household applications. These substances are non-toxic, non-flammable, and non-reactive with other chemical compounds. These desirable safety characteristics, along with their stable thermodynamic properties, make them ideal for many applications--as coolants for commercial and home refrigeration units, aerosol propellants, electronic cleaning solvents, and blowing agents.
- **Halons** are fire-fighting agents that were introduced into Australia in the early 1970s. They quickly replaced many previously accepted fire-fighting products because of their superior fire-fighting characteristics and ease of use.Halons are fully halogenated chemicals that have relatively long lifetimes in the atmosphere. They are broken down in the stratosphere releasing reactive bromine that is extremely damaging to ozone. Reactions involving bromine are estimated to be responsible for 25 per cent of the chemical destruction of ozone over Antarctica and 50 per cent over the Arctic. The ozone depleting potential of halons is 10 times greater than that of chloroflurocarbons (CFCs). As such, halons are a very aggressive ozone depleting chemical. One kilogram of halon 1211 can destroy 50 tonnes of ozone.
- **Methyl Bromide** is an ozone depleting substance (ODS) used as:
    - a soil fumigant in horticultural industries
    - a pest control treatment on dry commodities such as packaged rice
    - a quarantine and pre-shipment (QPS) treatment for imports, exports and certain commodities transported interstate
    - as a feedstock in chemical reactions to create other chemicals
- **Carbon tetrachloride (CCl_4),** despite its toxicity, was first used in the early 1900s as a fire extinguishant, and more recently as an industrial solvent, an agricultural fumigant, and in many other industrial processes including petrochemical refining, and pesticide and pharmaceuticals

production. Recently it has also been used in the production of CFC-11 and CFC-12. It has accounted for less than 8% of total ozone depletion.

- **Methyl chloroform,** also known as 1, 1, 1 trichloroethane is a versatile, all-purpose industrial solvent used primarily to clean metal and electronic parts. It was introduced in the 1950s as a substitute for carbon tetrachloride. Methyl chloroform has accounted for roughly 5% of total ozone depletion.

6.7 DEFORESTATION AND DESERTIFICATION

6.7.1 Deforestation

Forests are vanishing at a rate of some 17 million hectares per year, an area about half the size of Finland. A minimum of 140 plant and animal species are Condemned to extinction each day. 58.8% of Indian forests have no or inadequate regeneration, mainly because of degradation. Consequently, the productivity of India's forests is very low averaging 0.7 cu. m hectare, per year. This is only a fourth of the Asian average. Livestock currently requires 932 million tonnes green and 780 million tonnes dry fodder annually, yet, only 2~O and 414 million tonnes respectively are provided. The rest come from grazing in forest land. Over 90 million livestock graze in India's forests, which currently have the ability to sustainably support only 31 million. There is an urgent need to grow fodder to protect the cattle wealth, the backbone of rural economy. Fuel wood consumption is estimated at 235 million tonnes per annum .While annual production is only 90 million tonnes, 40 million tonnes from forests and the remaining from agro wastes. The balance comes from forest capital stock. India has among the lowest per capita availability of forests in the world of 0.11 ha, as compared to 0.50 in Thailand and 0.13 in China. Industrial wood consumption is 28 million cu .m per annum ,but annual forest growth is only 122 million cu.m.The remainder is made up by depletion of forests beyond their regenerative capacities.

6.7.2 Desertification

The processes by which an area becomes even more barren, less capable of retaining vegetation, and progresses towards becoming a desert. This is often a cause of long term disasters. This may result either due to a natural phenomenon linked to climatic change or due to abusive land use. In fact even for climatic change, these are the improper land use practices which are largely responsible. Removal of vegetal cover brings about marked changes in the local climate of the area. Thus, deforestation, overgrazing etc. bring about changes in rainfall, temperature, wind velocity etc. and also lead to soil erosion. Such changes then lead to desertification of the area. Desertification often starts as patchy destruction of productive land. Increased dust particles in atmosphere lead to desertification and drought in margins of the zones that are not humid. Even the humid zones are in danger of getting progressively drier if droughts continue to occur over a series of years. Indications are clear that the temporary phenomena of meteorological drought in India is tending to become permanent one. This trend is not restricted to the fringes of existing deserts only. The threat of desertification is thus real, because a s the forest diminishes, there is steady rise in the atmospheric temperature.

There is no denying that during the past two decades, there has been increased damage to forests and other ecosystems by man. At the time of independence in the country, out of 75 million hectares, about 22% was under forest cover. Today this has been reduced to 19%. India has been losing 10 million trees every 24 years. Thus, deforestation is one of the main factors leading to desertification, primarily through its effect on climate of the area. As a result of gross mismanagement of natural resources including land, certain irreversible changes have triggered the breakdown of nutrient cycles and microclimatic equilibrium in the soil indicating the onset of desertic conditions.

The impact of drought on the socio-cultural milieu is significant. The most obvious adverse impact is felt on the health of cattle and human population of the affected area. Water scarcity results in a higher incidence of water diseases. And while drought is a temporary phenomenon, desertification is not.

The chief causes of desertification are (i) climatic factors and (ii) human factors (human cultures). Recent changes in land use and population density had much ecological effects. Human factors are population growth, increased density, reduced nomadism and loss of grazing lands, and (iii) interactions between climate and culture.

6.8 INTERNATIONAL CONVENTIONS / PROTOCOLS

6.8.1 Earth summit

The issues addressed in Earth Summit are :

- systematic scrutiny of patterns of production particularly the production of toxic components, such as lead in gasoline, or poisonous waste including radioactive chemicals
- alternative sources of energy to replace the use of fossil fuels which are linked to global climate change

- new reliance on public transportation systems in order to reduce vehicle emissions, congestion in cities and the health problems caused by polluted air and smog
- the growing scarcity of water

The Convention on Biological Diversity was opened for signature at the Earth Summit, and made a start towards redefinition of money supply measures that did not inherently encourage destruction of natural ecoregions and so-called uneconomic growth. The Earth Summit resulted in the following documents:

- *Rio Declaration on Environment and Development*
- *Agenda*
- *Convention on Biological Diversity*
- *Forest Principles*
- *Framework Convention on Climate Change (UNFCCC).*

Both *Convention on Biological Diversity* and *Framework Convention on Climate Change* were set as legally binding agreements.

6.8.2 Kyoto Protocol

The Kyoto Protocol is a protocol to the United Nations Framework Convention on Climate Change (UNFCCC or FCCC), aimed at fighting global warming. The UNFCCC is an international environmental treaty with the goal of achieving stabilization of greenhouse gas concentrations in the atmosphere at a level that would minimize dangerous anthropogenic interference with the climate system.

Under the Protocol, 37 industrialized countries called as Annex 1 countries. commit themselves to a reduction of four greenhouse gases (GHG) namely carbon dioxide, methane, nitrous oxide, sulphur hexafluoride and two groups of gases like hydrofluorocarbons and perfluorocarbons produced by them, and all member countries give general commitments. Annex I countries agreed to reduce their collective greenhouse gas emissions by 5.2% from the 1990 level. Emission limits do not include emissions by international aviation and shipping, but are in addition to the industrial gases, chlorofluorocarbons, or (CFCs), which are dealt with under the 1987 Montreal Protocol on Substances that Deplete the Ozone Layer.

The Protocol allows for several flexible mechanisms, such as emissions trading, the Clean Development Mechanism (CDM) and joint implementation to allow Annex I countries to meet their GHG emission limitations by purchasing GHG emission reductions credits from elsewhere, through financial exchanges, projects that reduce emissions in non-Annex I countries, from other Annex I countries, or from annex I countries with excess allowances. Kyoto is intended to cut global emissions of greenhouse gases.

The objective is the stabilization and reconstruction of greenhouse gas concentrations in the atmosphere at a level that would prevent dangerous anthropogenic interference with the climate system. The objective of the Kyoto climate change conference was to establish a legally binding international agreement, whereby all the participating nations commit themselves to tackling the issue of global warming and greenhouse gas emissions. The target agreed upon was an average reduction of 5.2% from 1990 levels by the year 2012. Contrary to popular belief, the Protocol will not expire in 2012. In 2012, Annex I countries must have fulfilled their obligations of reduction of greenhouse gases emissions established for the first commitment period (2008–2012). The five principal concepts of the Kyoto Protocol are:

- commitments to reduce greenhouse gases that are legally binding for annex I countries, as well as general commitments for all member countries.
- implementation to meet the Protocol objectives, to prepare policies and measures which reduce greenhouse gases, increasing absorption of these gases (for example through geosequestration and biosequestration) and use all mechanisms available, such as joint implementation, clean development mechanism and emissions trading; being rewarded with credits which allow more greenhouse gas emissions at home.
- minimizing impacts on developing countries by establishing an adaptation fund for climate change.
- accounting, reporting and review to ensure the integrity of the Protocol.
- compliance by establishing a compliance committee to enforce commitment to the Protocol.

6.8.3 Montreal Protocol

The Montreal Protocol on Substances That Deplete the Ozone Layer ,a protocol to the Vienna Convention for the Protection of the Ozone Layer is an international treaty designed to protect the ozone layer by phasing out the production of a number of substances believed to be responsible for ozone depletion. The treaty was opened for signature on September 16, 1987, and entered into force on January 1, 1989, followed by a first meeting in Helsinki, May 1989. Since then, it has undergone seven revisions, in 1990 (London), 1991 (Nairobi), 1992 (Copenhagen), 1993 (Bangkok), 1995

(Vienna), 1997 (Montreal), and 1999 (Beijing). It is believed that if the international agreement is adhered to, the ozone layer is expected to recover by 2050. Due to its widespread adoption and implementation it has been hailed as an example of exceptional international co-operation with Kofi Annan quoted as saying that "perhaps the single most successful international agreement to date has been the Montreal Protocol". It has been ratified by 196 states.

REVIEW QUESTIONS

1. Write a note on Greenhouse gasses and greenhouse effect.
2. Explain the causes of global warming.
3. Explain the impacts of climate change on human environment.
4. Discuss the concept of Ozone depletion and Ozone depleting substances.
5. Write short notes on deforestation and desertification.
6. Write notes on (i) Kyoto protocol (ii) Montreal protocol.

CHAPTER 7

ENVIRONMENTAL POLICY, LEGISLATION, RULES AND REGULATIONS

7.1 NATIONAL ENVIRONMENTAL POLICY ACT (NEPA)

NEPA Contains three important sections. They are:

(i) The declaration of national environmental policies and goals. (ii) The establishment of action-forcing provisions for federal agencies to enforce those policies and goals and (iii) The establishment of a Council on Environmental Quality (CEQ) in the Executive Office of the President.

The essential purpose of NEPA is to ensure that environmental factors weighted equally when compared to other factors in the decision making process undertaken by federal agencies. The act establishes the national environmental policy, including a multidisciplinary approach to considering environmental effects in federal government agency decision making. The act also established the President's Council on Environmental Quality (CEQ). The CEQ was established to advise the President in the preparation of an annual environmental quality report addressing the state of federal agencies in implementing the act, on national policies nurture and promote the improvement of the environments quality and on the state of the environment. The effectiveness of NEPA originates in its requirement of federal agencies to prepare an environmental statement to accompany reports and recommendations for funding from Congress. This document is called an Environmental Impact Statement (EIS). NEPA is an action-forcing piece of legislation, meaning that the act itself does not carry any criminal or civil sanctions. All enforcement of NEPA was to be obtained through the process of the court system.

A major federal action has been expanded to include most things that a federal agency could prohibit or regulate. In practice, a project is required to meet NEPA guidelines when a federal agency provides any portion of the financing for the project. Sometimes, however, review of a project by a federal employee can be viewed as a federal action and would then, therefore, require NEPA-compliant analysis be performed.

NEPA covers a vast array of federal agency actions, but not all actions are covered necessarily covered under NEPA. The act does not apply to purely private or purely public state action. This means that there is a complete absence of government influence or funding concerning that specific action. Exemptions and exclusions are also present within NEPA's guidelines. Exemptions from NEPA include specific federal projects detailed in legislation, EPA exemptions and functional equivalent exemptions. A functional Equivalent exemption applies where compliance with other environmental laws requires environmental analysis similar to NEPA. These other environmental laws can include but are not limited to the Clean Air Act, Resource Conservation and Recovery Act, Safe Drinking Water Act and the Federal Insecticide, Fungicide and Rodenticide Act. The last exclusion form NEPA is a categorical exclusion. These exclusions are simply consisting of categories of actions which do not individually or cumulatively have a significant effect on the human environment.

7.2 ENVIRONMENTAL (PROTECTION) ACT

The Environment (Protection) Act (EPA), 1986 not only has important constitutional implications but also an international background. The spirit of the proclamation adopted by the United Nations Conference on Human Environment, held in Stockholm in June 1972, was implemented by the Government of India by creating this Act. Although there were several existing laws that dealt directly or indirectly with environmental issues it was necessary to have a general legislation for environmental protection because the existing laws focused on very specific types of pollution, or specific categories of hazardous substances, or were indirectly related to the environment through laws that control land use, protect our national parks and sanctuaries and our wildlife. However, there was no over arching legislation and certain areas of environmental hazards were not covered. There were also gaps in areas that were potential environmental hazards and there were essentially related to the multiplicity of regulatory agencies. Thus, there was a need for an authority to study, plan and implement the long-term requirements of environmental safety, and direct and coordinate a system of appropriate response to emergencies threatening the environment. This Act was thus passed to protect the environment, as there was a growing concern over the deteriorating state of the environment. As impacts grew considerably environmental protection became a national priority in the 1970s. While the wider general legislation to protect our environment is now in place, it has become increasingly evident that our environmental situation continues to deteriorate. We need to implement this Act much more aggressively if our environment is to be protected. The presence of excessive concentrations of harmful chemicals in the atmosphere and aquatic ecosystems leads to the disruption of food chains and a loss of species. Pubic concern and support is crucial for implementing the EPA. This must be supported by an enlightened media, good administrators, highly aware policy makers, informed judiciary and trained technocrats who together can influence and prevent further degradation of our environment. Each of us has a responsibility to make this happen.

7.3 LEGAL ASPECTS AIR (PREVENTION AND CONTROL OF POLLUTION) ACT, 1981

It is also a comprehensive legislation with more than fifty sections. It makes provisions, inter alia, for Central and State Boards, power to declare pollution control areas, restrictions on certain industrial units, authority of the Boards to limit emission of air pollutants, power of entry, inspection, taking samples and analysis, penalties, offences by companies and Government and cognizance of offences etc.

The Air Act is largely a replica of the Water Act. With elaborate provisions prescribing the powers and functions of the Central and State Air Boards. The Act specifically empowers State Government to designate air pollution areas and to prescribe the type of fuel to be used in these designated areas. According to this Act, no person can operate certain types of industries including the asbestos, cement, fertilizer and petroleum industries without consent of the State Board. The Board can predicate its consent upon the fulfillment of certain conditions. The Air Act apparently adopts an industry wide "best available technology" requirement. As in the Water Act, courts may hear complaints under the Act only at the instigation of, or with the sanction of, the State Board. J.

7.4 WATER (PREVENTION AND CONTROL OF POLLUTION) ACT, 1974

The Water Act 1974 with certain amendments in 1978 is an extensive legislation with more than sixty sections for the prevention and control of water pollution. Among other things, the Act provides for constitution of central and state boards for preventing water pollution, power to take water samples and their analysis, discharge of sewage or trade effluents, appeals, revision, minimum and maximum penalties, publication of names of offenders, offences by companies and Government departments, cognizance of offences, water laboratories, analysis etc. Prevention and control of water pollution is achieved through a permit or 'consent administration' procedure. Discharge of effluents is permitted by obtaining the consent of the State Water Board, subject to any condition they specify. Any person who fails to comply with a directive of the State cannot, however, entertain in suit under this Act unless the suit is brought by, or with the sanction of the State Board.

7.5 WATER POLLUTION CESS ACT 1977

According to this Act, anyone consuming water has to pay certain amount of cess depending on 1. Whether the industry is using water for industrial cooling, spraying in mine pits or boilers feed, 2. For domestic purposes, 3. In processing, whereby water gets polluted and pollutants are easily biodegradable, and 4. In processing whereby water gets polluted and the pollutants are not easily bio-degradable and are toxic. Those industries that had installed a suitable treatment plant for the treatment of industrial effluents can get a rebate of 70 per cent on the cess payable.

Appreciating the desirability of wildlife conservation the Government of India enacted the Wildlife (Protection) Act, 1972. The act, a landmark in the history of wildlife legislation in our country by which wildlife was transferred from State list to concurrent list in 1976, thus giving power to the Central Government to enact the legislation. In India, nearly 134 animal species have been regarded as threatened. As against this 1 mammalian and 3 avian species have already become extinct, 71 mammals, 47 birds and 15 reptile species have been declared threatened. A National Wildlife action plan has been prepared whose objective is to establish a network of scientifically managed areas such as national parks, sanctuaries and biosphere reserves, to cover representative and viable samples of all significant bio-geographic subdivisions within the country. Thus, as of today, there are 67 national parks and 394 sanctuaries in India, representing 4 per cent of the geographic area of the country. The major activities and provisions in the act can be summed up as follows:

1. It defines the wildlife related terminology.
2. It provides for the appointment of wildlife advisory board, wildlife warden, their powers, duties etc. Under the Act, comprehensive listing of endangered wildlife species was done for the first time and prohibition of hunting of the endangered species was mentioned.
3. Protection to some endangered plants like Bed dome cycad, Blue Vanda,
4. Ladies Slipper Orchid, Pitcher plant etc. is also provided under the Act.
5. The act provides for setting up of National Parks, Wild life Sanctuaries etc.
6. The Act provides for the constitution of Central Zoo Authority.
7. There is provision for trade and commerce in some wildlife species with license for sale, possession, transfer etc.
8. The Act imposes a ban on the trade or commerce in scheduled animals.
9. It provides for legal powers to officers and punishment of offenders.
10. It provides for captive breeding programme for endangered species. Several conservation projects for individual endangered species like lion (1972), tiger (1973), crocodile (1974), and brown antlered deer (1981) were started under this Act. The Act is adopted by all states in India except J&K, which has its own Act!

7.6 FOREST CONSERVATION ACT

To appreciate the importance of the Forest Conservation Act of 1980, which was amended in 1988, it is essential to understand its historical background. The Indian Forest Act of 1927 consolidated all the previous laws regarding forests that were passed before the 1920s. The Act gave the Government and Forest Department the power to create Reserved Forests, and the right to use Reserved Forests for Government use alone. It also created protected forests, in which the use of resources by local people was controlled. Some forests were to be controlled by the village community, and these were called village Forests. The Act remained in force till the 1980s when it was realized that protecting forests for timber production alone was not acceptable. The other values of protecting the services that forests provide and its valuable assets such as biodiversity began to overshadow the importance of their revenue earnings from timber. Thus a new Act was essential. This led to the Forest Conservation Act of 1980 and its amendment 1988. India's first Forest Policy was enunciated in 1952. Between 1952 and 1988, the extent of deforestation was so great that it became essential to formulate a new policy on forests and their utilization. The earlier forest policies had focused only on revenue generation. In the 1980's it became clear that forests must be protected for their other functions such as the maintenance of soil and water regimes centered around ecological concerns. It also provided for the use of goods and services of the forest for its local inhabitants. The new policy framework made conversion of forests into other uses much less possible. Conservation of the forests as a natural heritage finds a place in the new policy, which includes the preservation of its biological diversity and genetic resources. It also values meeting the needs of local people for food, fuel wood, fodder and Non Timber Forest Produce or NTFPs. It gives priority to maintaining environmental stability and ecological balances. It expressly states that the network of protected areas should be strengthened and extended.

In 1992, the 73rd and 74th Amendments to the Constitution furthered governance through panchayats. They allow the states to invest the local panchayats with the authority to manage local forest resources. The Forest Conservation Act of 1980 was enacted to control deforestation. It ensured that forestlands could not be de-reserved without prior approval of the Central Government. This was created as some states had begun to de reserve the reserved forests for non-forest use. These states had regularized encroachments and resettled 'project affected people' from development projects such as dams in these de-reserved areas. The need for a new legislation became urgent. The Act made it possible to retain a greater control over the frightening level of deforestation in the country and specified penalties for offenders. Penalties for offences in Protected Forests: A person who commits any of the following offences like felling trees, stripping the bark or leaves of trees, setting fire to such forests, kindling a fire without taking precautions to prevent its spreading, dragging timber, or permitting cattle to damage any tree, shall be punishable with imprisonment for a term which may extend to six months or with a fine which may extend to Rs. 500, or both. When there is a reason to believe that a forest offence has been committed pertaining to any forest produce, the produce together with all tools used in committing such offences may be seized by any forest officer or police officer. Every officer seizing any property under this section shall put on the property a mark indicating the seizure and report the seizure to the Magistrate who has the jurisdiction to try the offence. Any forest officer, even without an order from the Magistrate or a warrant, can arrest any person against whom a reasonable suspicion exists.

7.7 MUNCIPAL SOLID WASTE MANAGEMENT AND HANDLING RULES

7.7.1 Functional Elements and Municipal Solid Waste Management

The activities associated with the management of municipal solid wastes from the point of generation to the final disposal can be grouped into six functional elements viz., waste generation, waste handling, sorting, storage, and processing at the source, waste collection, waste sorting, processing and transformation, waste transfer and transport and waste disposal. Current thinking on the best methods to deal with waste is centered on a broadly accepted "Hierarchy of waste management" (arrangement in order of rank), which gives a priority listing of waste management options available (Figure 1). The hierarchy gives important general guidelines on the relative desirability of different management options. The hierarchy usually adopted is (a) waste minimization/reduction at source, (b) recycling, (c) waste processing (with recovery of resources i.e. materials (products) and energy), (d) waste transformation without recovery of resources and (e) disposal on land (land filling).

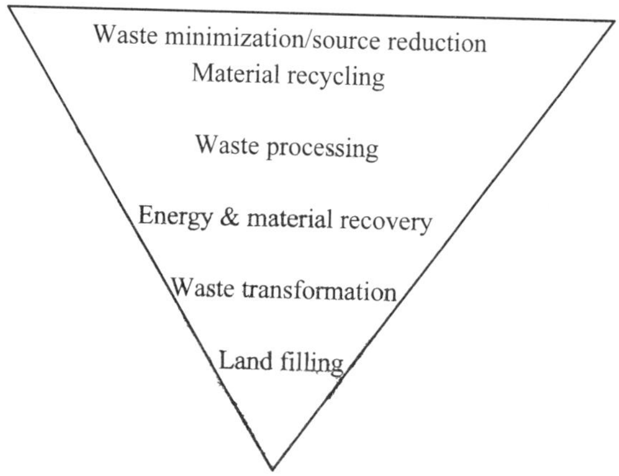

Fig. 7.1 Hierarchy of waste management.

The highest rank of the MSWM hierarchy is waste minimization or reduction at source, which involves reducing the amount of the wastes produced. Reduction of the waste at the source is first in the hierarchy because it is the most effective way to reduce the quantity of waste, the cost associated with it and its environmental impacts.

7.7.2 Collection of municipal solid wastes

1. Littering of municipal solid waste shall be prohibited in cities, towns and in urban areas notified by the State Governments. To prohibit littering and facilitate compliance, the following steps shall be taken by the municipal authority, namely :-
   - (i) Organising house-to-house collection of municipal solid wastes through any of the methods, like community bin collection (central bin), house-to-house collection, collection on regular pre-informed timings and scheduling by using bell ringing of musical vehicle (without exceeding permissible noise levels);
   - (ii) Devising collection of waste from slums and squatter areas or localities including hotels, restaurants, office complexes and commercial areas;
   - (iii) Wastes from slaughter houses, meat and fish markets, fruits and vegetable markets, which are biodegradable in nature, shall be managed to make use of such wastes;
   - (iv) Bio-medical wastes and industrial wastes shall not be mixed with municipal solid wastes and such wastes shall follow the rules separately specified for the purpose;

(v) Collected waste from residential and other areas shall be transferred to community bin by hand-driven containerised carts or other small vehicles;

(vi) Horticlutural and construction or demolition wastes or debris shall be separately collected and disposed off following proper norms. Similarly, wastes generated at dairies shall be regulated in accordance with the State laws;

(vii) Waste (garbage, dry leaves) shall not be burnt;

(viii) Stray animals shall not be allowed to move around waste storage facilities or at any other place in the city or town and shall be managed in accordance with the State laws.

2. The municipal authority shall notify waste collection schedule and the likely method to be adopted for public benefit in a city or town.

3. It shall be the responsibility of generator of wastes to avoid littering and ensure delivery of wastes in accordance with the collection and segregation system to be notified by the municipal authority as per para 1(2) of this Schedule.

7.7.3 Segregation of municipal solid wastes

In order to encourage the citizens, municipal authority shall organise awareness programmes for segregation of wastes and shall promote recycling or reuse of segregated materials. The municipal authority shall undertake phased programme to ensure community participation in waste segregation. For this purpose, regular meetings at quarterly intervals shall be arranged by the municipal authorities with representatives of local resident welfare associations and non-governmental organizations.

7.7.4 Storage of municipal solid wastes

Municipal authorities shall establish and maintain storage facilities in such a manner as they do not create unhygienic and insanitary conditions around it. Following criteria shall be taken into account while establishing and maintaining storage facilities, namely :-

(i) Storage facilities shall be created and established by taking into account quantities of waste generation in a given area and the population densities. A storage facility shall be so placed that it is accessible to users;

(ii) Storage facilities to be set up by municipal authorities or any other agency shall be so designed that wastes stored are not exposed to open atmosphere and shall be aesthetically acceptable and user-friendly;

(iii) Storage facilities or 'bins' shall have 'easy to operate' design for handling, transfer and transportation of waste. Bins for storage of bio-degradable wastes shall be painted green, those for storage of recyclable wastes shall be printed white and those for storage of other wastes shall be printed black;

(iv) Manual handling of waste shall be prohibited. If unavoidable due to constraints, manual handling shall be carried out under proper precaution with due care for safety of workers.

7.7.5 Transportation of municipal solid wastes

Vehicles used for transportation of wastes shall be covered. Waste should not be visible to public, nor exposed to open environment preventing their scattering. The following criteria shall be met, namely:-

(i) The storage facilities set up by municipal authorities shall be daily attended for clearing of wastes. The bins or containers wherever placed shall be cleaned before they start overflowing;

(ii) Transportation vehicles shall be so designed that multiple handling of wastes, prior to final disposal, is avoided.

7.7.6 Processing of municipal solid wastes

Municipal authorities shall adopt suitable technology or combination of such technologies to make use of wastes so as to minimize burden on landfill. Following criteria shall be adopted, namely:-

(i) The biodegradable wastes shall be processed by composting, vermicomposting, anaerobic digestion or any other appropriate biological processing for stabilization of wastes. It shall be ensured that compost or any other end product shall comply with standards as specified in Schedule-IV;

(ii) Mixed waste containing recoverable resources shall follow the route of recycling. Incineration with or without energy recovery including pelletisation can also be used for processing wastes in specific cases. Municipal authority or the operator of a facility wishing to use other state-of-the-art technologies shall approach the Central Pollution Control Board to get the standards laid down before applying for grant of authorisation.

7.7.7 Disposal of municipal solid wastes

Land filling shall be restricted to non-biodegradable, inert waste and other waste that are not suitable either for recycling or for biological processing. Land filling shall also be carried out for residues of waste processing facilities as well as pre-processing rejects from waste processing facilities. Land filling of mixed waste shall be avoided unless the same is found unsuitable for waste processing. Under unavoidable circumstances or till installation of alternate facilities, land-filling shall be done following proper norms. Landfill sites shall meet the specifications as given in Schedule –III.

7.7.8 Specifications for Landfill Sites

Site Selection

In areas falling under the jurisdiction of 'Development Authorities' it shall be the responsibility of such Development Authorities to identify the landfill sites and hand over the sites to the concerned municipal authority for development, operation and maintenance. Elsewhere, this responsibility shall lie with the concerned municipal authority.

Selection of landfill sites shall be based on examination of environmental issues. The Department of Urban Development of the State or the Union territory shall co-ordinate with the concerned organisations for obtaining the necessary approvals and clearances.

The landfill site shall be planned and designed with proper documentation of a phased construction plan as well as a closure plan.

The landfill sites shall be selected to make use of nearby wastes processing facility. Otherwise, wastes processing facility shall be planned as an integral part of the landfill site.

The existing landfill sites which continue to be used for more than five years, shall be improved in accordance of the specifications given in this Schedule.

Biomedical wastes shall be disposed off in accordance with the Bio-medical Wastes (Management and Handling) Rules, 1998 and hazardous wastes shall be managed in accordance with the Hazardous Wastes (Management and Handling) Rules, 1989, as amended from time to time.

The landfill site shall be large enough to last for 20-25 years.

The landfill site shall be away from habitation clusters, forest areas, water bodies monuments, National Parks, Wetlands and places of important cultural, historical or religious interest.

A buffer zone of no-development shall be maintained around landfill site and shall be incorporated in the Town Planning Department's land-use plans.

Landfill site shall be away from airport including airbase. Necessary approval of airport or airbase authorities prior to the setting up of the landfill site shall be obtained in cases where the site is to be located within 20 km of an airport or airbase..

Facilities at the Site

Landfill site shall be fenced or hedged and provided with proper gate to monitor incoming vehicles or other modes of transportation.

The landfill site shall be well protected to prevent entry of unauthorised persons and stray animals.

Approach and other internal roads for free movement of vehicles and other machinery shall exist at the landfill site.

The landfill site shall have wastes inspection facility to monitor wastes brought in for landfill, office facility for record keeping and shelter for keeping equipment and machinery including pollution monitoring equipments.

Provisions like weigh bridge to measure quantity of waste brought at landfill site, fire protection equipments and other facilities as may be required shall be provided.

Utilities such as drinking water (preferably bathing facilities for workers) and lighting arrangements for easy landfill operations when carried out in night hours shall be provided.

Safety provisions including health inspections of workers at landfill site shall be periodically made.

Specifications for land filling

Wastes subjected to land filling shall be compacted in thin layers using landfill compactors to achieve high denslly of the wastes. In high rainfall areas where heavy compactors cannot be used alternative measures shall be adopted.

Wastes shall be covered immediately or at the end of each working day with minimum 10 cm of soil, inert debris or construction material till such time waste processing facilities for composting or recycling or energy recovery are set up as per Schedule I.

Prior to the commencement of monsoon season, an intermediate cover of 40-65 cm thickness of soil shall be placed on the landfill with proper compaction and grading to prevent infiltration during monsoon. Proper drainage berms shall be constructed to divert run-off away from the active cell of the landfill.

After completion of landfill, a final cover shall be designed to minimize infiltration and erosion. The final cover shall meet the following specifications, namely :-

(a) The final cover shall have a barrier soil layer comprising of 60 cms of clay or amended soil with permeability coefficient less that 1×10^{-7} cm/sec.

(b) On top of the barrier soil layer there shall be a drainage layer of 15 cm.

(c) On top of the drainage layer there shall be a vegetative layer of 45 cm to support natural plant growth and to minimize erosion.

7.7.9 Pollution Prevention

In order to prevent pollution problems from landfill operations, the following provisions shall be made, namely :-

(a) Diversion of storm water drains to minimize leachate generation and prevent pollution of surface water and also for avoiding flooding and creation of marshy conditions;

(b) Construction of a non-permeable lining system at the base and walls of waste disposal area. For landfill receiving residues of waste processing facilities or mixed waste or waste having contamination of hazardous materials (such as aerosols, bleaches, polishes, batteries, waste oils, paint products and pesticides) minimum liner specifications shall be a composite barrier having 1.5 mm high density polyethylene (HDPE) geomembrane, or equivalent, overlying 90 cm of soil (clay or amended soil) having permeability coefficient not greater than 1 x 10-7 cm/sec. The highest level of water table shall be at least two meter below the base of clay or amended soil barrier layer;

(c) Provisions for management of leachates collection and treatment shall be made. The treated leachates shall meet the standards specified in Schedule- IV;

(d) Prevention of run-off from landfill area entering any stream, river, lake or pond.

7.7.9.1 Water Quality Monitoring

Before establishing any landfill site, baseline data of ground water quality in the area shall be collected and kept in record for future reference. The ground water quality within 50 metres of the periphery of landfill site shall be periodically monitored to ensure that the ground water is not contaminated beyond acceptable limit as decided by the Ground Water Board or the State Board or the Committee. Such monitoring shall be carried out to cover different seasons in a year that is, summer, monsoon and post-monsoon period.

Usage of groundwater in and around landfill sites for any purpose (including drinking and irrigation) is to be considered after ensuring its quality. The following specifications for drinking water quality shall apply for monitoring purpose, namely :-

Table 7.1

S.No	Parameters	IS 10500: 1991 Desirable limit (mg/l except for pH)
1.	Arsenic	0.05
2.	Cadmium	0.01
3.	Chromium	0.05
4.	Copper	0.05
5.	Cyanide	0.05
6.	Lead	0.05
7.	Mercury	0.001
8.	Nickel	-
9.	Nitrate as NO_3	45.0
10.	PH	6.5-8.5
11.	Iron	0.3
12.	Total hardness (as $CaCO_3$)	300.0
13.	Chlorides	250
14.	Dissolved solids	500
15.	Phenolic compounds (as C6H5OH)	0.011
16	Zinc	5.0
17.	Sulphate (as SO_4)	200

7.7.9.2 Ambient Air Quality Monitoring

Installation of landfill gas control system including gas collection system shall be made at landfill site to minimize odour generation, prevent off-site migration of gases and to protect vegetation planted on the rehabilitated landfill surface.

The concentration of methane gas generated at landfill site shall not exceed 25 per cent of the lower explosive limit (LEL).

The landfill gas from the collection facility at a landfill site shall be utilized for either direct thermal applications or power generation, as per viability. Otherwise, landfill gas shall be burnt (flared) and shall not be allowed to directly escape to the atmosphere or for illegal tapping. Passive venting shall be allowed if its utiliztion or flaring is not possible.

Ambient air quality at the landfill site and at the vicinity shall be monitored to meet the following specified standards, namely :-

Table 7.2

S.No	Parameters	Acceptable levels
(i)	Sulphur dioxide	120 $\mu\gamma/m^3$ (24 hours)
(ii)	Suspended Particulate Matter	500 $\mu\gamma/m^3$ (24 hours)
(iii)	Methane	Not to exceed 25 per cent of the lower explosive limit (equivalent to 650 $\mu\gamma/m^3$)
(iv)	Ammonia daily average	
	(Sample duration 24 hrs)	0.4 $\mu\gamma/m^3$ (400 $\mu\gamma/m^3$)
(v)	Carbon monoxide	1 hour average : 2 $\mu\gamma/m^3$ 8 hour average : 1 $\mu\gamma/m^3$

The ambient air quality monitoring shall be carried out by the concerned authority as per the following schedule, namely:-

(a) Six times in a year for cities having population of more than fifty lakhs;

(b) Four times in a year for cities having population between ten and fifty lakhs;

(c) Two times in a year for town or cities having population between one and ten lakhs.

7.7.9.3 Plantation at Landfill Site

A vegetative cover shall be provided over the completed site in accordance with the and following specifications, namely :-

(a) Selection of locally adopted non-edible perennial plants that are resistant to drought and extreme temperatures shall be allowed to grow;

(b) The plants grown be such that their roots do not penetrate more than 30 cms. This condition shall apply till the landfill is stabilised;

(c) Selected plants shall have ability to thrive on low-nutrient soil with minimum nutrient addition;

(d) Plantation to be made in sufficient density to minimize soil erosion.

7.7.9.4 Closure of Landfill Site and Post-care

The post-closure care of landfill site shall be conducted for at least fifteen years and long term monitoring or care plan shall consist of the following, namely :

(a) Maintaining the integrity and effectiveness of final cover, making repairs and preventing run-on and run-off from eroding or otherwise damaging the final cover;

(b) Monitoring leachate collection system in accordance with the requirement;

(c) Monitoring of ground water in accordance with requirements and maintaining ground water quality;

(d) Maintaining and operating the landfill gas collection system to meet the standards.

Use of closed landfill sites after fifteen years of post-closure monitoring can be considered for human settlement or otherwise only after ensuring that gaseous and leachate analysis comply with the specified standards.

7.7.9.5 Special Provisions for Hilly Areas

Cities and towns located on hills shall have location-specific methods evolved for final disposal of solid wastes by the municipal authority with the approval of the concerned State Board or the Committee. The municipal authority shall set up processing facilities for utilization of biodegradable organic wastes. The inert and non-biodegradable waste shall be used for building roads or filling-up of appropriate areas on hills. Because of constraints in finding adequate land in hilly areas, wastes not suitable for road-laying or filling up shall be disposed of in specially designed landfills.

7.7.10 Standards for Composting, Treated Leachates and Incineration

1. The waste processing or disposal facilities shall include composting, incineration, pelletisation, energy recovery or any other facility based on state-of-the-art technology duly approved by the Central Pollution Control Board

2. In case of engagement of private agency by the municipal authority, a specific agreement between the municipal authority and the private agency shall be made particularly, for supply of solid waste and other relevant terms and conditions.

3. In order to prevent pollution problems from compost plant and other processing facilities, the following shall be complied with, namely :-

    (i) The incoming wastes at site shall be maintained prior to further processing. To the extent possible, the waste storage area should be covered. If, such storage is done in an open area, it shall be provided with impermeable base with facility for collection of leachate and surface water run-off into lined drains leading to a leachate treatment and disposal facility;

    (ii) Necessary precautions shall be taken to minimise nuisance of odour, flies, rodents, bird menace and fire hazard;

    (iii) In case of breakdown or maintenance of plant, waste intake shall be stopped and arrangements be worked out for diversion of wastes to the landfill site;

    (iv) Pre-process and post-process rejects shall be removed from the processing facility on regular basis and shall not be allowed to pile at the site. Recyclables shall be routed through appropriate vendors. The non-recyclables shall be sent for well designed landfill site(s).

    (v) In case of compost plant, the windrow area shall be provided with impermeable base. Such a base shall be made of concrete or compacted clay, 50 cm thick, having permeability coefficient less than 10^{-7} cm/sec. The base shall be provided with 1 to 2 per cent slope and circled by lined drains for collection of leachate or surface run-off;

    (vi) Ambient air quality monitoring shall be regularly carried out particularly for checking odour nuisance at down-wind direction on the boundary of processing plant.

In order to ensure safe application of compost, the following specifications Table 7.3 for compost quality shall be met, namely:

Table 7.3

Parameters	Concentration not to exceed * (mg/kg dry basis, except pH value and C/N ratio)
Arsenic	10.00
Cadmium	5.00
Chromium	50.00
Copper	300.00
Lead	100.00
Mercury	0.15
Nickel	50.00
Zinc	1000.00
C/N ratio	20-40
PH	5.5-8.5

* Compost (final product) exceeding the above stated concentration limits shall not be used for food crops. However, it may be utilized for purposes other than growing food crops.

4. The disposal of treated leachates shall follow the following standards, namely:

Table 7.4

S.No	Parameter	Standards (Mode of Disposal)		
		Inland surface water	Public sewers	Land disposal
1.	Suspended solids, mg/l, max	100	600	200
2.	Dissolved solids (inorganic) mg/l, max	2100	2100	2100
3.	PH value	5.5 to 9.0	5.5 to 9.0	5.5 to 9.0
4.	Ammonical nitrogen (as N), mg/l, max.	50	50	-
5.	Total Kjeldahl nitrogen (as N), mg/l, max.	100	-	-
6.	Biochemical oxygen demand (3 days at 27 ^{0}C) max.(mg/l)	30	350	100
7.	Chemical oxygen demand, mg/l, max.	250	-	-
8.	Arsenic (as As), mg/l, max	0.2	0.2	0.2
9.	Mercury (as Hg), mg/l, max	0.01	0.01	-
10.	Lead (as Pb), mg/l, max	0.1	1.0	-
11.	Cadmium (as Cd), mg/l, max	2.0	1.0	-
12.	Total Chromium (as Cr), mg/l, max.	2.0	2.0	-
13.	Copper (as Cu), mg/l, max.	3.0	3.0	-
14.	Zinc (as Zn), mg/l, max.	5.0	15	-
15.	Nickel (as Ni), mg/l, max	3.0	3.0	-
16.	Cyanide (as CN), mg/l, max.	0.2	2.0	0.2
17.	Chloride (as Cl), mg/l, max.	1000	1000	600
18.	Fluoride (as F), mg/l, max	2.0	1.5	-
19	Phenolic compounds (as C_6H_5OH) mg/l, max.	1.0	5.0	-

Note : While discharging treated leachates into inland surface waters, quantity of leachates being discharged and the quantity of dilution water available in the receiving water body shall be given due consideration

The incinerators shall meet the following operating and emission standards, namely:

7.7.11 Operating Standards

1. The combustion efficiency (CE) shall be at least 99.00%.
2. The combustion efficiency is computed as follows :

$$C.E. = \frac{\%CO_2}{\%CO_2 + \%CO} \times 100$$

1. Emission Standards

	Parameters	Concentration mg/Nm^3 at (12% CO_2 correction)
1.	Particulate matter	150
2.	Nitrogen Oxides	450
3.	HCl	50
4.	Minimum stack height shall be 30 metres above ground.	
5.	Volatile organic compounds in ash shall not be more than	0.01%.

Note :

1. Suitably designed pollution control devices shall be installed or retrofitted with the incinerator to achieve the above emission limits, if necessary.
2. astes to be incinerated shall not be chemically treated with any chlorinated disinfectants
3. Chlorinated plastics shall not be incinerated.
4. Toxic metals in incineration ash shall be limited within the regulatory quantities as specified in the Hazardous Wastes (Management and Handling) Rules, 1989 as amended from time to time.
5. Only low sulphur fuel like l.d.o., l.s.h.s or Diesel shall be used as fuel in the incinerator.

7.8 BIOMEDICAL WASTE MANAGEMENT AND HANDLING RULES

Hospital waste is generated during the diagnosis, treatment, or immunization of human beings or animals or in research activities in these fields or in the production or testing of biologicals. It may include wastes like sharps, soiled waste, disposables, anatomical waste, cultures, discarded medicines, chemical wastes, etc. These are in the form of disposable syringes, swabs, bandages, body fluids, human excreta, etc. This waste is highly infectious and can be a serious threat to human health if not managed in a scientific and discriminate manner. It has been roughly estimated that of the 4 kg of waste generated in a hospital at least 1 kg would be infected. These wastes are categorized into 10 different categories as:

1. Human anatomical waste (tissues, organs, body parts etc.)
2. Animal waste
3. Microbiology and biotechnology waste, such as, laboratory cultures, microorganisms, human and animal cell cultures, toxins etc.
4. Waste sharps such as, hypodermic needles, syringes, scalpels, broken glass etc
5. Discarded medicines and cyto-toxic drugs
6. Soiled waste, such as dressings, bandages, plaster casts, material contaminated with blood etc
7. Solid waste (disposal items like tubes, catheters etc., excluding sharps)
8. Liquid waste generated from any of the infected areas
9. Incineration ash
10. Chemical waste

Surveys carried out by various agencies show that the health care establishments in India are not giving due attention to their waste management. After the notification of the Bio-medical Waste (Handling and Management) Rules, 1998, these establishments are slowly streamlining the process of waste segregation, collection, treatment, and disposal. Many of the larger hospitals have either installed the treatment facilities or are in the process of doing so.

7.8.1 Treatment and Disposal

1. Bio-medical waste shall be treated and disposed of in accordance with Schedule I, and in compliance with the standards prescribed in Schedule V.

2. Every occupier, where required, shall set up in accordance with the time-schedule in Schedule VI, requisite bio-medical waste treatment facilities like incinerator, autoclave, microwave system for the treatment of waste, or, ensure requisite treatment of waste at a common waste treatment facility or any other waste treatment facility.

Segregation, Packaging, Transportation and Storage

1. Bio-medical waste shall not be mixed with other wastes.
2. Bio-medical waste shall be segregated into containers/bags at the point of generation in accordance with Schedule II prior to its storage, transportation, treatment and disposal. The containers shall be labeled according to Schedule III.
3. If a container is transported from the premises where bio-medical waste is generated to any waste treatment facility outside the premises, the container shall, apart from the label prescribed in Schedule III, also carry information prescribed in Schedule IV.
4. Notwithstanding anything contained in the Motor Vehicles Act, 1988, or rules thereunder, untreated biomedical waste shall be transported only in such vehicle as may be authorised for the purpose by the competent authority as specified by the government.
5. No untreated bio-medical waste shall be kept stored beyond a period of 48 hours

Provided that if for any reason it becomes necessary to store the waste beyond such period, the authorised person must take permission of the prescribed authority and take measures to ensure that the waste does not adversely affect human health and the environment.

Prescribed Authority

1. The Government of every State and Union Territory shall establish a prescribed authority with such members as may be specified for granting authorisation and implementing these rules. If the prescribed authority comprises of more than one member, a chairperson for the authority shall be designated.
2. The prescribed authority for the State or Union Territory shall be appointed within one month of the coming into force of these rules.
3. The prescribed authority shall function under the supervision and control of the respective Government of the State or Union Territory.
4. The prescribed authority shall on receipt of Form 1 make such enquiry as it deems fit and if it is satisfied that the applicant possesses the necessary capacity to handle bio-medical waste in accordance with these rules, grant or renew an authorisation as the case may be.

5. An authorisation shall be granted for a period of three years, including an initial trial period of one year from the date of issue. Thereafter, an application shall be made by the occupier/operator for renewal. All such subsequent authorisation shall be for a period of three years. A provisional authorisation will be granted for the trial period, to enable the occupier/operator to demonstrate the capacity of the facility.
6. The prescribed authority may after giving reasonable opportunity of being heard to the applicant and for reasons thereof to be recorded in writing, refuse to grant or renew authorisation.
7. Every application for authorisation shall be disposed of by the prescribed authority within ninety days from the date of receipt of the application.
8. The prescribed authority may cancel or suspend an authorisation, if for reasons, to be recorded in writing, the occupier/operator has failed to comply with any provision of the Act or these rules :

Provided that no authorisation shall be cancelled or suspended without giving a reasonable opportunity to the occupier/operator of being heard.

Authorisation

1. Every occupier of an institution generating, collecting, receiving, storing, transporting, treating, disposing and/or handling bio-medical waste in any other manner, except such occupier of clinics, dispensaries, pathological laboratories, blood banks providing treatment/service to less than 1000 (one thousand) patients per month, shall make an application in Form 1 to the prescribed authority for grant of authorisation.
2. Every operator of a bio-medical waste facility shall make an application in Form 1 to the prescribed authority for grant of authorisation.
3. Every application in Form 1 for grant of authorisation shall be accompanied by a fee as may be prescribed by the Government of the State or Union Territory.

Advisory Committee

The Government of every State/Union Territory shall constitute an advisory committee. The committee will include experts in the field of medical and health, animal husbandry and veterinary sciences, environmental management, municipal administration, and any other related department or organisation including non-governmental organisations. The State

Pollution Control Board/Pollution Control Committee shall be represented. As and when required, the committee shall advise the Government of the State/Union Territory and the prescribed authority about matters related to the implementation of these rules.

Annual Report

Every occupier/operator shall submit an annual report to the prescribed authority in Form 11 by 31 January every year, to include information about the categories and quantities of bio-medical wastes handled during the preceding year. The prescribed authority shall send this information in a compiled form to the Central Pollution Control Board by 31 March every year.

Maintenance of Records

1. Every authorised person shall maintain records related to the generation, collect ' ion, reception, storage, transporation, treatment, disposal and/or any form of handling of bio-medical waste in accordance with these rules and any guidelines issued.
2. All records shall be subject to inspection and verification by the prescribed authority at any time.

Accident Reporting

When any accident occurs at any institution or facility or any other site where bio-medical waste is handled or during transportation of such waste, the authorised person shall report the accident in Form III to the prescribed authority forthwith.

Appeal

Any person aggrieved by an order made by the prescribed authority under these rules may, within thirty days from the date on which the order is communicated to him, prefer an appeal to such authority as the Government of State/Union Territory may think fit to constitute :

Provided that the authority may entertain the appeal after the expiry of the said period of thirty days if it is satisfied that the appellant was prevented by sufficient cause from filing the appeal in time.

Categories of Bio-Medical Waste

As per Ministry of Environment and Forest (MoEF) all the biomedical waste is categorized as below (Table 7.5).

Table 7.5

Option	Waste Category	Treatment & Disposal
Category No. I	Human Anatomical Waste (human tissues, organs, body parts)	incineration@/deep burial*
Category No. 2	Animal Waste (animal tissues, organs, body parts carcasses, bleeding parts, fluid, blood and experimental animals used in research, waste generated by veterinary hospitals colleges, discharge from hospitals, animal houses)	incineration@/deep burial*
Category No 3	Microbiology & Biotechnology Waste (wastes from laboratory cultures, stocks or specimens of micro-organisms live or attenuated vaccines, human and animal cell culture used in research and infectious agents from research and industrial laboratories, wastes from production of biologicals, toxins, dishes and devices used for transfer of cultures)	local autoclaving/micro-waving/incineration@
Category No 4	Waste sharps (needles, syringes, scalpels, blades, glass, etc. that may cause puncture and cuts. This includes both used and unused sharps)	disinfection (chemical treat-ment@01/auto claving/micro-waving and mutilation/ shredding"
Category No 5	Discarded Medicines and Cytotoxic drugs (wastes comprising of outdated, contaminated and discarded medicines)	inc ineratio n@/destruct ion and drugs disposal in secured landfills
Category No 6	Solid Waste (Items contaminated with blood, and body fluids including cotton, dressings, soiled plaster casts, lines, beddings, other material contaminated with blood)	incineration@ autoclaving/microwaving

Table 7.2 *Contd...*

Option	Waste Category	Treatment & Disposal
Category No. 7	Solid Waste (wastes generated from disposable items other than the waste shaprs such as tubings, catheters, intravenous sets etc).	disinfection by chemical treatment@@ autoclaving/ microwaving and mutilation/ shredding##
Category No. 8	Liquid Waste (waste generated from laboratory and washing, cleaning, house-keeping and disinfecting activities)	disinfection by chemical treatment@@ and discharge into drains.
Category No. 9	Incineration Ash (ash from incineration of any bio-medical waste)	disposal in municipal landfill
Category No. 10	Chemical Waste (chemicals used in production of biologicals, chemicals used in disinfection, as insecticides, etc.)	chemical treatment@@ and discharge into drains for liquids and secured landfill for solids

@@ Chemicals treatment using at least 1% hypochlorite solution or any other equivalent chemical reagent. It must be ensured that chemical treatment ensures disinfection.

Multilation/shredding must be such so as to prevent unauthorised reuse.

@ There will be no chemical pretreatment before incineration. Chlorinated plastics shall not be incinerated.

* Deep burial shall be an option available only in towns with population less than five lakhs and in rural areas.

Table 7.6 Colour coding and type of container for disposal of bio-medical wastes

Colour Conding	Type of Container -I Waste Category	Treatment options as per Schedule I
Yellow	Plastic bag Cat. 1, Cat. 2, and Cat. 3, Cat. 6.	Incineration/deep burial
Red	Disinfected container/plastic bag Cat. 3, Cat. 6, Cat.7.	Autoclaving/Microwaving/ Chemical Treatment
Blue/White translucent	Plastic bag/puncture proof Cat. 4, Cat. 7. Container	Autoclaving/Microwaving/ Chemical Treatment and destruction/shredding
Black	Plastic bag Cat. 5 and Cat. 9 and Cat. 10. (solid)	Disposal in secured landfill

Notes:

1. Colour coding of waste categories with multiple treatment options as defined in Schedule I, shall be selected depending on treatment option chosen, which shall be as specified in Schedule I.
2. Waste collection bags for waste types needing incineration shall not be made of chlorinated plastics.
3. Categories 8 and 10 (liquid) do not require containers/bags.
4. Category 3 if disinfected locally need not be put in containers/bags.

7.9 HAZARDOUS WASTE MANAGEMENT AND HANDLING RULES

Hazardous wastes may be defined as wastes of industrial, institutional or consumer origin which because of their physical, chemical or biological Characteristics are potentially dangerous to human and the environment. In Some cases although the active agents may be liquid or gaseous, they are classified as solid waste because they are confined in solid containers. Typical Examples are solvents, paints and pesticides whose spent containers are frequently mixed with municipal wastes and become part of urban waste tream. Certain hazardous wastes cause explosions in incinerators and fires at landfill sites. Others, such as pathological wastes room hospitals and radioactive wastes, require special handling at all time. Good management practice should ensure that hazardous wastes are stored, collected, transported and disposed off separately, preferably after suitable treatment to render them innocuous.

7.9.1 Categories of Hazardous Wastes

Waste Categories	Typc of wastes	Regulatory Quantities
Waste Category No. 4	Mercury, Arsenic, Thallium and Cadmium bearing wastes.	5 kilogrammes per year the sum of the specified substance calculated as pure metal.
Waste Category No. 5	Non-halogenated hydrocarbons including solvent.	200 kilogrammes per year calculated as non-halogenated hydrocarbons
Waste Category No. 6	Halogenated hydro-carbon including solvents	50 kilograms per year calculated as helogenated hydrocarbons.
Waste Category No. 7	Wastes from paints, pigments, glue, varnish and printing ink.	250 kilogrammes per year calculated as oil or oil emulsions.
Waste Category No.8	Wastes from Dyes and Dye intermediate containing inorganic chemical compounds.	200 kilogrammes per year calculated as inorganic chemicals.

Contd....

Waste Categories	Type of wastes	Regulatory Quantities
Waste Category No. 9	Wastes from Dyes and Dye intermediate containing organic chemical compounds.	50 kilogrammes per year calculated as organic chemicals.
Waste Category No. 10	Waste oil and oil emulsions.	1000 kilogrammes per year calculated as oil and oil emulsions.
Waste Category No. 11	Tarry wastes from refining and tar residues from distillation or prolytic treatment.	200 kilogrammes per year calculated as tar
Waste Category No. 12	Sludgesarising from treatment of waste waters containing heavy metals, toxic organics, oils emulsions and spend chemical and inceneration ash.	irrespective of any quantity.
Waste Category No. 13	Phenols.	5 kilogrammes per year calculated as phenols.
Waste Category No. 14	Asbestos.	200 kilogrammes per year calculated asbestos.
Waste Category No. 15	Wastes from manufacturing of pesticides and herbicides and residues from pesticides and, herbicides formulation units	5 kilogrammes per year calculated as pesticides and their intermediate products.
Waste Category No. 16	Acid/Alkaline/Slurry	200 kilogrammes per year calculated as Acids/Alkalies.
Wastes Category No. 17	Off-specification and discarded products.	Irrespective of any quantity.
Wastes Category No.18	Discarded containers and Containers linears of hazardous and toxic wastes.	Irrespective of any quantity.

REVIEW QUESTIONS

1. Write the three important sections of National Environment Policy Act (NEPA) of USA and explain.
2. Write short notes on (i) Environment protection Act (ii) Air act (iii) Water act (iv) Water pollution costs act (v) Forest conservation act
3. Discuss the municipal solid waste management and handling rules.
4. Discuss the biomedical waste management rules.
5. Discuss the Hazardous waste management rules.
6. Listout the various categories of hazardous wastes.
7. What do you understand segregation of solid waste?
8. What are the rules of storage of municipal solid wastes?
9. What are the technologies for processing of municipal solid wastes?

CHAPTER 8

ENVIRONMENTAL IMPACT ASSESSMENT

8.1 INTRODUCTION

Environmental Impact Assessment (EIA) is a study of the probable environmental impacts of a development project. Environment protection act of 1986 requires such studies prior to proceeding with any project in India. Effect on the natural environment caused by human actions. Includes indirect effects, for example, through pollution, as well as direct effects such as cutting down trees.

The purpose of EIA is to identify potential environmental impacts from proposals, such as projects and programs, and to propose means to avoid or reduce the significant impacts. EIA was developed formally in the 1970s and has been incorporated in the procedures of governments and major development organizations world-wide. As a result it is important that people who have any role in the design or planning of projects, or may be associated with deciding about their suitability, should be aware of EIA and how it operates in their local area. The Ministry of Environment and Forests of India have been in a great effort in environmental impact assessment in India. The main laws in nation are water act (1974), The Indian wildlife (protection) act (1972), The air (prevention and control of pollution) act (1981) and the environment (protection) act (1986). The responsible bodies are Central Pollution Control Board (CPCB), State Pollution Control Boards (SPCB) and state environmental impact assessment authorities.

Environmental Impact Assessment (EIA) studies need a significant amount of primary and secondary environmental data. The primary data are those which need to be collected in the field to define the status of environment like air quality data, water quality data biological data, etc. The secondary data are those data which have been collected over the years and can be used to understand the existing environmental scenario of the project area. The environmental impact assessment studies are conducted over a short period of time and therefore understanding the environmental trends based on few months of primary data has its own limitations. Ideally, the primary data has to be considered along with the secondary data for complete understanding of the existing environmental status of the area.

Environmental impact assessment is systematic process used to help make the decisions that take in to account the impacts of human activities on the environment. The process helps to include the environmental factors in to the project proposal. The EIA involves an assessment of the potential impacts of the project on the ecosystem. In EIA, existing data related to the environment i.e. base line, data is collected. Then predictions are made about the emissions. from the proposed projects. These predicted values are superimposed on the existing values using different models to predict the level of pollution after the project commences. Thus, the future values help to make the decision whether the proposed project would be: environmentally' sustainable in the region or not. Since other sustainability factors are also considered in EIA, it gives a complete picture of a region after the implementation of the project.

8.2 DEFINITION AND SCOPE OF EIA

Environmental Impact Assessment (EIA) can be defined as the systematic identification and evaluation of the potential impacts (effects) of proposed projects, plans, programs or legislative actions relative to the physical, chemical, biological, cultural and socio-economic components of the total environment. The primary purpose of the EIA process is to encourage the consideration of the environmental issues in planning and decision making and to ultimately arrive at actions which are more environmentally compatible (L. Canty, 1996).

The potential scope of a comprehensive EIA system is considerable and could include the appraisal of policies, plans, programmes and specific projects. EIA, as it has developed in many countries, involves a number of procedures and stages:

- Identification of projects requiring EIA, sometimes known as screening;
- Identification of the key issues to be addressed in an EIA, called scoping;
- Impact assessment and evaluation;
- Impact mitigation and monitoring;
- Review of the completed Environmental Impact Statement and;
- Public participation.

The result of an EIA is assembled in a document known as an Environmental Impact Statement (EIS) which looks at all the positive and negative effects of a particular project on the environment. This report is just one component of the information required to aid decision makers in making their ultimate decision about a project. EIA can be considered as a mechanism which maximizes the efficient use of natural and human resources. It can also reduce costs and time taken to reach a decision by ensuring that subjectivity and duplication of effort are minimized, as well as identifying and attempting to evaluate the primary and secondary consequences which might require the introduction of expensive pollution control equipment or compensation and other costs at a later date.

"Environment" in EIA context mainly focuses, physical, chemical, biological, geological, social, economical, and aesthetic dimensions along with .their complex interactions, which affects individuals, communities and ultimately determines their forms, character, relationship, and survival. Sustainable development is built on three basic pillars: economic growth, ecological balance and social progress. Economic growth achieved in a way that does not consider, the environmental concerns, will not be sustainable in the long run. However, sustainable development needs careful integration of three constituents—environmental, economic, and social needs in order to achieve both an increased standard of living in the short term, and a net gain or equilibrium among human, natural, and economic resources to support future generations in the long term. "It is necessary to understand the links between environment and development in order to make development choices that will be economically efficient, socially equitable and responsible, and environmentally sound." EIA have two roles: legal and educational.

- The legal one is quite straight forward to ensure that development projects such as a housing estate, a road/bridge or some such construction project has a minimal impact on the environment in its entire 'lifecycle' - i.e. during design, construction, use, maintenance, and demolition. Many countries now have laws stipulating that unless an EIA study is carried out particularly for large infrastructure projects, permission for construction will not be granted by the local authority.
- The educational role is equally important to educate everyone involved both professionals and users. We need to look at all our daily actions as eventually and cumulatively affecting the environment. This includes our daily choices, where a delicate balance between financial and environmental considerations needs to be made automatically without thinking.

There is a need to emphasize and to promote the new paradigm of preventive the management to protect the environment in the process of development of the nation. Various tools of preventive management be classified into following three groups. (Table 8.1)

Table 8.1

Management based tools	Process based tools	Product based tools
Environmental management system	Environmental technology Assessment	Industrial ecology
Environmental performance evaluation	Toxic use reduction	Extended producers Responsibility
Environmental audits	Best operating practices	Eco-labeling
Environmental reporting and communication	Environmentally best practice	Design for environment
Total cost accounting	Best available technology	Life cycle assessment
Law and policy	Pollution prevention	
Trade and environment	Cleaner production	
Environmental economics	Cleaner technology	
	Eco-efficiency	

Indicators can be classified in to environmental performance indicators and environmental condition indicators. The environmental performance indicators can be further divided into two categories namely. operational performance indicators and management performance indicators. The operational performance indicators are related to the process and other operational activities of the organization, and would typically address the issue of raw material consumption, energy consumption, water consumption in the organization, the quantities of waste water generated, other solid wastes generated, emission from the organization.

8.3 CLASSIFICATION OF EIA

EIA can be classified based on the purpose and the theme of development. EIA can be climate impact assessment, demographic impact assessment, development impact assessment, ecological impact assessment, economic and fiscal impact assessment, health impact assessment, risk assessment, social impact assessment, strategic impact assessment, technology assessment.

In addition to this list, EIA is also categorised based on systematic analysis of environmental parameters, geographical region, carrying capacity limitations and sectoral planning. They are strategic EIA, regional EIA, sectoral EIA, project level EIA and life cycle assessment.

8.3.1 Strategic EIA (SEIA)

Strategic EIA refers to systematic analysis of the environmental effects of development policies, plans, programs and other proposed strategic actions. This process extends the aims and principles of EIA upstream in the decision- making process, beyond the project level and when major alternatives are still open. Strategic EIA represents a proactive approach to integrate environmental considerations into the higher level of decision-making.

8.3.2 Regional EIA

EIA in the context of regional planning integrates environmental concerns into development planning for a geographic region, normally at the sub-country level. Such an approach is referred to as the economic-cum-environmental (EcE) development planning. This approach facilitates adequate integration of economic development with management of renewable natural resources within the carrying capacity limitation to achieve sustainable development. It fulfils the need for macro-level environmental integration, which the project-oriented EIA is unable to address effectively. Regional EIA addresses the environmental impacts of regional development plans and thus, the context for project-level EIA of the subsequent projects, within the region. In addition, if environmental effects are considered at regional level, then cumulative environmental effects of all the projects within the region can be accounted.

8.3.3 Sectoral EIA

Instead of project-level-EIA, an EIA should take place in the context of regional and sectoral level planning. Once sectoral level development plans have the integrated sectoral environmental concerns addressed, the scope of project-level EIA will be quite narrow. Sectoral EIA will help to address specific environmental problems that may be encountered in planning and implementing sectoral development projects.

8.3.4 Project Level EIA

Project level EIA refers to the developmental activity in isolation and the impacts that it exerts on the receiving environment. Thus, it may not effectively integrate the cumulative effects of the development in a region.

8.3.5 Life Cycle Assessment

A broader approach to deal with environmental impacts in manufacturing is called life cycle analysis. This approach recognizes that environmental concerns enter into every step of the process with respect to the manufacturing, of the products and thus examines environmental impacts of the product at all stages of the product life cycle. This includes the product design, development, manufacturing, packaging, distribution, usage and disposal. LCA is concerned with reducing environmental impacts at all these stages and looking at the. total picture rather than just one stage of the production process. Through utilizing this concept, firms minimize the life cycle environmental costs of their total product system. LCA gives sufficient scope to think about the. alternatives, which are lower at cost.

From the above discussion, it is clear that EIA shall be integrated at all the levels i.e. strategic, regional, sectoral and the project level. Whereas, the strategic EIA is a structural change, the regional EIA refers to the substantial information processing and drawing complex inferences. The project-level EIA is relatively simple and reaches to meaningful conclusions. As we progress and the resource planning concepts emerge in our decision- making process, the integration of overall regional issues will become part of the impact assessment studies.

8.4 GROUPING OF ENVIRONMENTAL IMPACTS

Environmental impacts resulting from proposed actions can be grouped into the following categories:

- Beneficial or detrimental
- Naturally reversible or irreversible
- Repairable via management practices or irreparable
- Short term or long term
- Temporary or continuous
- Occurring during construction phase or operational phase
- Local, regional, national or global
- Accidental or planned (recognized before hand)
- Direct (primary) or Indirect (secondary)
- Cumulative or single

Based on characteristics of impacts, the impacts could 'fall within three broad classifications namely direct, indirect and cumulative. The assessment of direct, indirect and cumulative impacts should not be considered in isolation or considered as separate stages in the EIA. Ideally, the assessment of such impacts should form an integral part of all stages of the EIA.

8.4.1 Direct Impacts

Direct impacts occur through direct interaction of an activity with an environmental, social, or economic component. For example, a discharge of any industry or an effluent from the Effluent Treatment Plant (ETP) from the industrial estates into a river may lead to a decline in water quality in terms of high biological oxygen demand (BOD) or dissolved oxygen (DO) or rise of water toxins.

8.4.2 Indirect Impacts

Indirect impacts on the environment are these which are not a direct result of the project, often produced away from or as a result of a complex impact pathway. The indirect impacts are also known as secondary or even third level impacts. For example, ambient air SO_2 rise due to stack emissions may deposit on land as SO_4 and cause acidic soils. Another example of indirect impact is the decline in water quality due to rise in temperature of water bodies receiving cooling water discharge from the nearby industry. This may, in turn, lead to a secondary indirect impact on aquatic flora in that water body and may further cause reduction in fish population. Reduction in fishing harvests, affecting the income of fishermen is a third level impact. Such impacts are characterized as socioeconomic (third level) impacts. The indirect impacts may also include growth- inducing impacts and other effects related to induced changes to the pattern of land use or additional road network, population density or growth rate (e.g. around a power project). In the process, air, water and other natural systems including the ecosystem may also be affected.

8.4.3 Cumulative Impacts

Cumulative impact consists of an impact that is created as a result of the combination of the project evaluated in the EIA together with other projects causing related impacts. These impacts occur when the incremental impact of the project is combined with the cumulative effects of other past, present and reasonably foreseeable future projects.

8.4.4 Induced Impacts

The cumulative impacts can be, due to induced actions of projects and activities that may occur if. the action under assessment is implemented such as growth inducing impacts and other effects related to induced changes to the pattern of future land use or additional road network, population density or growth rate. Induced actions may not be officially announced or be part of any official plan. Increase in workforce and nearby communities contributes to this effect. They usually have no direct relationship with the action under assessment and represent the growth-inducing potential of an action. New roads leading from those constructed for a project, increased recreational activities, and construction of new service facilities are examples of induce actions.

However, the cumulative impacts due to induced development or third level or even secondary indirect impacts are difficult to be quantified. Because of higher levels' of uncertainties, these' impacts cannot be normally assessed over a long time horizon. An EIA practitioner usually can only guess as to what such induced impacts may be and the possible extent of their implications on the environmental factors.

8.5 BASELINE DATA ACQUISITION

To conduct EIA study for any given developmental project, plan and legislative action, the baseline data of the project area of 10 Km radial buffer with project site as its centre is required. The acquisition of such data can be carried out through environmental inventory.

Rapid urbanization has resulted in a significant impact on the environment. Now all the concerned authorities have realized that development cannot be sustained unless the environment is protected. Therefore, the Governmental authorities in most of the countries in the world have made it mandatory for project developers, plan makers, action implementation organization and for the legislative action to carryout EIA of their developmental activity or project and prepare Environmental Management Plans (EMP) so that the environmental quality is protected. In order to assess the likely impacts arising out of the project, action or any plan, number of various environmental parameters are necessary and need to be collected, derived from various sources and also to prepare the EMP for the proposed activity.

Three of the most significant terms are environmental inventory, environment impact assessment and Environmental Impact Statement (EIS). These three are interlinked and without inventory, assessment is not possible and without knowing the probable impacts, it is not possible to prepare EIS.

8.5.1 Environmental Inventory

"Environmental inventory" is a complete description of the environment as it exists in an area where a particular proposed action is being considered The inventory is compiled from a checklist of descriptors for the physical - chemical, biological, cultural, and socioeconomic environments. The "physical-chemical environment" includes such major areas as soils, geology topography surface-water and groundwater resources, water quality, air quality, and climatology. The "biological environment" refers to the flora and fauna of the area, including species of trees, grasses, fish, herpetofauna, birds, and mammals. Specific reference must be made to any threatened and/or endangered plant or animal species. General biological features such as species diversity and overall ecosystem stability should also be presented. Items in the "cultural environment" include historic and archaeological sites, and aesthetic resources such as visual quality. The "socioeconomic environment" refers to a range of considerations related to humans in the environment, including population trends and population distributions; economic indicators of human welfare; educational systems; transportation networks and other infrastructure concerns such as water supply, wastewater disposal, and solid-waste management; public services such as police and fire protection and medical facilities; and many others. The physical-chemical and biological environments can be referred to as the "natural environment," or the "biophysical environment," while the cultural and socioeconomic environments represent the "man-made environment."

The health impacts of projects, plans, programs, or policies should be considered in the decision-making process. Because of the importance of these concerns, particularly in developing countries, an environmental health impact assessment process has been proposed (WHO, 1987). For certain types of projects, such as nuclear power plants, it may be necessary to address psychological impacts on nearby residents ("Can Change Damage Your Mental Health," 1982).

The emphasis in environmental impact studies in the early 1970s was on the physical-chemical and biological environments; however, added attention was given to the cultural and socioeconomic environments as the decade progressed. One reason for the attention to the socioeconomic environment was the emphasis on secondary impacts mentioned in the NEPA guidelines issued by the Council on Environmental Quality in 1973. Attention to health and ecological risks is increasing and this trend is expected to accelerate.

The environmental inventory serves as the basis for evaluating the potential impacts on the environment, both beneficial and adverse, of a proposed action. It is included in an environmental impact statement (EIS). Development of the inventory represents an initial step in the environmental impact assessment process. The scope of the environmental inventory or baseline data acquisition includes a detailed characterization of the environment in an area of 10 Km radius around the porposed facility for environmental components viz, air, noise, water, land, ecology and socio-economic environment. Acquisition of baseline data is mainly: (a) to assess the present status of air, noise, water, land, ecology and socio-economic components of the environment in an area of 10 Km radius around the proposed site, which is mandatory. (b) To identify and quantify significant impacts of processing operations on environmental components. (c) To prepare Environmental Management Plan (EMP) with pollution control technologies to be adopted for mitigation of adverse impacts and site specific remedial measures and (d) To delineate future environmental quality monitoring programmes to be pursued by the proponents after commissioning the proposed project.

8.5.2 Methodology

An area of 10 Km radial buffer zone around the proposed project has to be studied for characterization of various environmental components derived through environmental inventory and prominent features from the boundary of the proposed unit are to be examined. The methodology adopted for the study is a three step procedure like:

(i) review of the baseline environmental status

(ii) identification, prediction and evaluation of the impacts of the proposed project on the environment and

(iii) environmental Management Plan (EMP) to ensure minimization of adverse and enhancement of beneficial environmental impacts.

Base Line Environmental Status: Review of existing environmental status consisted of data collection on various environmental components such as air, water, land use, ecology, socio-economics. The data can be collected from both primary and secondary sources. Primary data can be collected through an environmental survey of the study area appropriately based on the seasons. Data on ambient air quality, water quality, noise levels, soil quality and flora-fauna will be collected from representative locations in the subject area of 10 Km radius. Data from secondary sources will also be collected from the concerned organizations.

Identification and Prediction of Impacts: Various impacts likely to occur due to the proposed project on the surrounding environment are to be identified. These impacts are assessed for their significance based on the background environmental quality in the area and the magnitude of the impact. All components of the environment are considered while identifying and assessing the impacts. Simulation models may be used wherever possible to predict impacts quantitatively. For further text, reader can refer the subsequent sections.

Environmental Management Plan: Based on the impacts identified, an appropriate environmental management strategy has to be developed and presented in the form of an EMP. The EMP consists of the various policies, control measures, etc., for abatement of critical environmental impacts arising out of the proposed project. The EMP is very effective at the implementation stages of the project. To determine the effectiveness of the EMP, an environmental quality-monitoring programme to be pursued by the proponents, was suggested to ensure smooth execution of the EMP and to monitor the changes in the ambient environmental quality due to the proposed project.

8.5.3 Data Products

For conducting EIA study for obtaining environmental clearance of any development project, number of data products are necessary and needed to acquire from various sources, by applying various techniques to derive the secondary data products, through number of field surveys for primary data products. The detailed methods of acquiring all types of data products is beyond the scope of this book. However the tables 8.2 and 8.3 provide the baseline data products required for EIA study. The readers can understand these tables easily as they are self explanatory.

8.5.4 Environmental Baseline Monitoring (EBM)

The term 'baseline' refers to conditions exiting before development against which subsequent changes can be referenced. EP act typically specifies that an EIA report should contain an environmental inventory to study the existing environment that would be or might be affected directly or indirectly by a proposed projcct. Environmental Baseline Monitoring (EBM) is a very important state of EIA. In fact, one can say that for an EIA in India the EBM is the .center of gravity. On one hand EBM plays a very vital role in EIA and on the other hand it provides feedback about the actual environmental impacts of a project. EBM during the operational phase helps in judging the success of mitigation measures in protecting the environment. They are also used to ensure compliance with environmental standards, and to facilitate any needed project design or operational changes.

The existing environment is broadly defined to include the natural, cultural, socio-economic systems and their interrelationships. The intention is not to describe all baseline conditions, but to focus the collection and description of baseline data on those Values Environmental Components (VECs) that are important and are affected by the proposed project activities and should be included for further impact assessments and lead to appropriate decision-making. EBM studies are carried out to: identify environmental conditions which might influence project design decisions, for example, site layout, structural or operational characteristics, identify sensitive issues or areas requiring mitigation or compensation, provide input data to analytical models used for predicting effects and provide baseline data against which the results of future monitoring programs can be compared.

At this stage of EIA process, the EBM is primarily discussed in the context of first purpose wherein the feedback from EBM programs may be used to: determine the available assimilative capacity of different environmental components within the designated impact zone and whether more or less stringent mitigation measures are needed; and improve the predictive capability of EIAs. There are many institutional, scientific, quality control, and fiscal issues that must be addressed in the implementation of an environmental monitoring program. Careful consideration of these issues in the design and planning stages will help to avoid many of the pitfalls associated with environmental monitoring programs.

Monitoring refers to the collection of data through a series of repetitive measurements of environmental parameters. The environmental quality monitoring programme design depends on the monitoring objectives specified for the selected area of interest. The main types of monitoring activities are: baseline monitoring, effects monitoring and compliance monitoring.

Baseline Monitoring: The measurement of environmental parameters during a pre-project period for the purpose of determining the range of variation of the system and establishing reference points against which changes can be measured. This leads to the assessment of the possible (additional available) assimilative capacity of the environmental components in pre-project period with respect to standard or target level.

Effects Monitoring: The measurement of environmental parameters during project construction and implementation to detect changes which are attributable to the project to provide the necessary information to: verify the accuracy of EIA predictions; and determine the effectiveness of measures to mitigate adverse effects of projects on the environment. Feedback from environmental effect monitoring programs may be used to improve the predictive capability of EIAs and also determine whether more or less stringent mitigation measures are needed a

Compliance Monitoring: The periodic sampling or continuous measurement of environmental parameters to ensure that regulatory requirements and standard are being met. Compliance and effects monitoring occurs during the project construction, operation, and abandonment stages. The resources and institutional set-up should be available for the: monitoring at these stages. All large-scale construction projects will require some construction stage monitoring. To control the environmental hazards of construction as specified in the ETA, a monitoring program should be established to ensure that each mitigation measure is effectively implemented. There are numerous potential areas for monitoring during operations.

Before any field monitoring tasks are undertaken, there are many institutional, scientific, and fiscal issues that must be addressed in the implementation of an environmental monitoring program. Careful. consideration of these issues in the design and planning stages will help avoid many of the pitfalls associated with environmental monitoring programs. Although these issues are important but the discussions here are confined to the monitoring network design component.

8.5.5 Environmental Monitoring Network Design

At the outset of planning for an environmental monitoring network, the EIA team manager may not know exactly what should be monitored, when monitoring should begin, where it should monitor, which techniques should be employed, and who should take responsibility for its conduct. Because there are usually a number of objective decisions associated with network design to be made. It is important to start with an analysis of environmental issues. The scoping phase of EIA is designed to identify and focus the major issues. Scoping should provide a valuable source of information on the concerns that need to be addressed by the monitoring network design; These are project specific as well as specific to the environmental setting of the location where the project is proposed to be located.

Hence, the network designs are associated with questions like: What are the expected outputs of the monitoring activity? Which problems do we need to address to? etc. Defining the output will influence the design of the network and optimize the resources used for monitoring. It will also ensure that the network is specially designed to optimize the information on the problems at hand.

What to Monitor?

The question of what to monitor is associated with the identification of Valued Environmental Component (VEC). VECs are the qualities or quantities of environmental values the country desire to protect and preserve in our environment. The environmental values are what we ultimately are trying to protect or are striving for with respect to the environment. Examples of environmental values are contaminant-free fish, or sustainable fisheries in our endeavor of Ecology. The choice of VECs is related to the perceived significant impact on the project implementation on important environmental components. In general, the significance or importance of environmental components maybe based on:

- legal protection provided (for example, rare and endangered species);
- political or public concerns (for example, resource use conflicts and sustainable development);
- scientific judgment (for example, ecological importance); or
- commercial or economic importance.

The chosen environmental indicators must be: 1. measurable; 2. appropriate to the scale of disturbance/ contamination, 3. appropriate to the impact mechanism, 4. appropriate and proportional to temporal dynamics; 5. diagnostic; and 6. standardized; as well as have: 1. a low natural variability; 2. a broad applicability; and 3. an existing data series.

Where, How and How Many Times to Monitor?

These are the other components of Monitoring Network Design. These questions are best answered based on local field conditions, capacity and resources available prevailing legal and regulatory priorities etc. For this screening or reconnaissance Surveys of the study area also necessary. This may also include some simple inexpensive measurements and assimilative / dispersion modeling. The data will give some information on the prevailing special and temporal variations, and the general background air pollution in the area. The number of monitoring stations and the indicators to be measured at each station in the final permanent network may then be decided upon as based on the results of the screening study as well as on the knowledge of the sources of the proposed development and prevailing local environmental/meteorological conditions. The best possible definition of the air pollution problem, together with the analysis of the resources: personnel, budget and equipment available, represent the basis for the decision on the following questions:

- What spatial density (number) of sampling stations is required.
- Where should the stations be located?
- What kind of equipment should be used?
- What additional background information is needed?
  - Meteorology
  - Topography
  - population density
  - emission sources and emission rates
  - effects and impacts.
- How will the data be accessible, communicated?

8.5.6 Monitoring Stations

This normally means that for designing a monitoring programme in an (study) area which might have an impact, several monitoring stations are needed for characterizing the baseline conditions of the impacted area. When considering the location of individual samplers, it is essential that the data collected are representative for the location and type of area without the undue influence from the immediate surroundings. In any measurement point in the study area the total ambient concentration is the representative of natural background concentration, regional background, and the impact of existing large regional sources such as-industrial emissions and other industries

To obtain the information about the importance of these different contributions it is therefore necessary to locate monitoring stations so that they are representative for different impacts. In addition to the ambient pollution data, one would often need other data governing the variations such as meteorological data for air pollution, to identify and quantify the sources contributing to the measurements.. When considering the location of individual samplers, it is essential that the data collected are representative for the location and type of area without undue influence from the immediate surroundings. Database support for different environmental components are tabulated in the Tables 8.2 and 8.3.

Table 8.2 Physical Environment Components and Indicators

Environmental Component	Environmental Indicators
Climatic variables	- Rainfall patterns – mean, mode, seasonality - Temperature patterns - Extreme events - Climate change projections - Prevailing wind – direction, speed, anomalies - Stability conditions and mixing height
Geology	- Underlying rock type - Surgical material - Geologic structures (faults etc.) - Geologic resources (minerals, etc.)
Topography	- slope form - Landform and terrain analysis - Specific landform types
Coastal dynamic and morphology	- Wave patterns - Currents - Shoreline morphology – near shore, foreshore - Sediment – characteristics and transport
Soil	- Type and characteristics - Porosity and permeability - Sub-soil permeability - Run-off rate - Effective depth (inches/centimeters) - Inherent fertility - Suitability for method of sewage disposal
Drainage	- Surface hydrology - Contours of the area for flow direction - Drainage network - Rainfall runoff relationships - Hydrogeology - Groundwater characteristics – springs, etc.
Water quality	- Terrestrial – rivers, lakes, ponds, gullies - coastal
Air quality	- Ambient - Respirable - Airshed importance - Odour levels
Noise	
Hazardous waste	

Table 8.3 Baseline Data required for EIA study

Attributes	Sampling		Measurement Method	Remarks
	Network	Frequency		
A. Air				
- Meteorological - Wind speed - Wind direction - Dry bulb temperature - Wet bulb temperature - Relative humidity - Rainfall - Solar radiation - Cloud cover	Minimum 1 site in the project impact area requirements Other additional site(s) are require depending upon the model applied or site sensitivities	Min: 1 hrly observations from continuous records	Mechanical/automatic weather station Rain gauge As per IMD As per IMP	IS 5182 Part 1-20 Sit-specific primary data is essential Secondary data from IMD, New Delhi for the nearest IMD station
Pollutants - SPM - RPM - SO_2 - NO_2 - CO - H_2S* - NH*3 - HC* - Fluoride* - Pb* - VOC-PAH* - Mercury* (parameters are given in points of coverage for EIA studies based on nature of project, raw material & process technology, location nature/activities within of air	10 to 15 locations in the project impact area	24 hrly twice a week 8 hrly twice a week 24 hrly twice a week	- Gravimetric (High – Volume) - Gravimetric (High – Volume with Cyclone) - EPA Modified West & Gaeke method - Arsenite Modified Jacob & Hochheiser - NDIR technique - Methylene-blue - Nessler's Method - Infra Red analyzer - Specific Ion meter	**Monitoring Network** Minimum 2 locations in upwind side. more sites in downwind side/impact zone All the sensitive receptors need to be covered **Measurement Methods** As per CPCB standards for NAQM, 1994
B. Noise				
Hourly equipment noise levels	Same as for Air Pollution along with others Identifies in study area	At least one day continuous in each season on a working and non-working day	Instrument: Sensitive Noise level meter (preferably recording type)	Min: IS: 4954-1968 as adopted by CPCB

Table 8.3 *Contd....*

Hourly equivalent noise levels	In plant (1.5 m from machinery or high emission processes)	Same as above for day and night	Instrument: Noise level meter	CPCB / OSHA
Hourly equivalent levels	Highways (within 500 meters from the road edge)	Same as above for day and right	Instrument: Noise level meter	CPCB / IS: 4954-1968
Peak particle velocity	150-200m from blast site	Based on hourly observations	PPV meters	
C. Water				
Parameters for water quality Ph, temp, turbidity, magnesium, hardness, total alkalinity, chloride, sulphate, nitrite, fluoride, sodium, potassium salinity Total nitrogen, total phosphorous, DO, BOD, COD, Phenol Heavy metals Total coliforms, faecal coliforms Pyto plankton	Set of grab samples during pre and post-monsson for ground and surface water for the whole study zone. For lab. Analysis the samples should be preserved for transport safe	Diurnal and season wise	Samples for water quality should be collected and analyzed as per: IS: 2488 (Part 1-5) methods for sampling and testing of industrial effluents Standard method for examination of water and waste water analysis published by American Public Health Association. International standard practices for benthos and aquatic flora & fauna	
Zooplankton Fish & other aquatic flora & fauna (parameters are given in points of coverage for EIA studies based on nature of project, raw material & process technology, location nature/activities within of air basin)				

Table 8.3 *Contd....*

For surface Water Bodies				
- Total Carbon - p^H - Dissolved Oxygen - Biological Oxygen - Demand - Free NH_4 - Boron - Sodium Absorption ratio - Electrical Conductivity	Monitoring locations should include upstream, on site, downstream of proposed discharge point. Besides sampling should cover width of the river in case water quality modeling is proposed. Standard methodology for collection of surface water (BIS standard) At least one grab sample per location per season	Yield & impact on water sources to be measured during critical season River Stretch within project area be divided in grids (say 1 km length and 1/3 width) and samples should be from each grid at a time when the wastewater discharged by other sources of pollution is expected to be maximum	Samples for water quality should be collected and analyzed as per: IS: 2488 (Part 1-5) methods for sampling and testing of industrial effluents Standard methods for examination of water and wastewater analysis published by American Public Health Association.	Historical data should be collected from relevant offices such as central water commission, state and central ground water board, Irrigation dept.
Parameters for wastewater characterization				
Temp, colour, odour, turbidity, TSS, TDS - PH, alkalinity as $CaCO_3$, p value, M value, total hardness as $CaCO_3$, chloride as Cl, sulphate as SO_4, Nitrate as NO_3, Floride as F, Phosphate as PO_4, Chromium as Cr (Hexavalent, total Ammonical Nitrogen as N, TKN, % sodium, BOD at 20 C, COD, DO, total residual chlorine as Cl_2, oil and grease, sulphide phenolic compound	Implant source depending upon the different waste streams the parameters can be optimized Grab and composite sampling representing avg of different process operations as well as worst emission scenario should be represented	Different operational cycles as well as raw material variations should be reflected in the analysis	Samples for water quality should be collected and analyzed as per: IS: 2488 (Part 1-5) methods for sampling and testing of industrial effluents Standard methods for examination of water and wastewater analysis published by American Public Health Association.	All plant sources categorize as: Different process waste streams as well as run-off conditions - ETP wastewater - Domestic / sanitary wastewater

Table 8.3 *Contd....*

D. Land Environment				
Soil - Particle size distribution - Texture pH - Electrical conductivity - Cation exchange capacity - Alkali metals - Sodium Absorption Ratio (SAR) - Permeability - Porosity	One surface sample from each landfill and/or hazardous waste site (if applicable) and prime villages, (soil samples be collected as per BIS specifications) in the study area	Season-wise	Collected and analyzed as per soil analysis reference book, M.I. Jackson and soil analysis reference book by C.A. Black	The purpose of impact assessment on soil (land environment) is to assess the significant impacts due to leaching of wastes or accidental releases and contaminating
Land use / Land Scape				
- Location code - Total project area - Topography - Drainage (natural) - Cultivated, forest plantations, water bodies roads and settlements	Atleast 20 points along with plant boundary and general major land use categories in the study area.	Drainage once in the study period and land use categories from secondary data (local maps) and satellite imageries	- Global positioning system - Topo-sheets - Satellite Imageries - (1 : 25,000) - Satellite Imageries - (1 : 25, 000)	Drainage within the plant area and surrounding is very important for storm water impacts. From land use maps sensitive receptors (forests parks, mangroves etc.) can be identified
Solid Waste:				
Quantities: Based on waste generated from per unit production Per capita contribution Collection, transport and disposal system Process Waste Quality (oily, chemical, biological)	For green field unites it is based on secondary data base of earlier plants.	Process wise or activity wise for respective raw material used. Domestic waste depends upon the season also	Guidelines IS 9569 : 1980 IS 10477 : 1983 IS 12625 : 1989 IS 12647 : 1989 IS 12662 (PTI) 1989	

Table 8.3 *Contd....*

Quality				
General segregation into biological / organic/ inert / hazardous Loss on heating pH EC Calorific value, metals etc.	Grab and Composite samples	Process wise or activity wise for respective raw material used. Domestic waste depends upon the season also	Analysis IS 9334 : 1979 IS 9235 : 1979 IS 10158 : 1982	
Hazardous waste				
- Permeability and porosity - Moisture pH - Electrical conductivity - Loss on ignition - Phosphorous - Total nitrogen - Cation exchange capacity - Particle size distribution - Heavy metal - Ansonia - Flouride	Grab and Composite samples, Recyclable components have to be analyzed for the recycling requirements	Process wise or activity wise for respective raw material used	Analysis IS 9334 : 1979 IS 9235 : 1979 IS 10158 : 1982	Impacts of hazardous waste should be performed critically depending on the waste characteristics and place of discharge. For land disposal the guidelines should be followed and impacts of accidental releases should be assessed
E. Biological Environment Aquatic				
- Primary productivity - Aquatic weeds - Enumeration of - phytoplankton, zooplankton and benthos - Fisheries - Diversity indices - Trophic levels	Considering probable impact, sampling points and number of samples to be decided on established guidelines on ecological	Season changes are very important	Standards techniques (APHA et. Al. 1995, Rau and Wooten 1980) to beam followed for sampling and measurement	Seasonal sampling for aquatic biota One season for terrestrial biota, in addition to vegetation studies during monsoon season

Table 8.3 *Contd....*

- Rare and endangered species - Sanctuaries / closed areas / Coastal regulation zone (CRZ) - Terrestrial - Vegetation – species list, economic importance, forest produce, medicinal value - Importance value index (IVI) of trees - Wild animals Avifauna - Rare and endangered species - Sanctuaries / National park / Biosphere reserve	studies based on site eco-environment setting within 10/25 km radius from the proposed site Samples to collect from upstream and downstream of discharge point, nearby tributaries at down stream, and also from dug wells close to activity site For forest studies, chronic as well as short-term impacts should be analyzed warranting data on micro climate conditions			Preliminary assessment Microscopic analysis of plankton and meiobenthos, studies of macrofauna, aquatic vegetation and application of indices, viz. Shannon, similarity, dominance IVI etc Point quarter plot-less method (random sampling) for terrestrial vegetation survey. Secondary data to collect from Government offices, NGOs, published literature Plankton net Sediment dredge Depth sampler Microscope Field binocular
F. Socio Economic				
- Demographic structure - Infrastructure resource base - Economic resource base - Health status: Morbidity pattern - Cultural and aesthetic attributes	Socio-economic survey is based on proportionate, stratified and random sampling method	Different impacts occurs during construction and operational phases of the project	Primary data collection through R&R surveys (if require) or community survey are based on personal interviews and questionaire	Secondary data from census records, statistical hard books, toposheets, health records and relevant official records available with Govt. agencies

8.6 IMPACTS ON ENVIRONMENTAL COMPONENTS

The environmental significance assigned to each event can be tabulated with each of the significance scores for each of the criterion specified for each type of project. The level of environmental significance assigned to each individual component/even can be used to determine the overall level of environmental impact of the proposed activity. The difficulty faced in reaching such a decision is where each event is assigned widely differing levels of significance thus requiring a weighting of the environmental severity of the various events. It is suggested that consideration should be given to the sensitivity of each event to the overall environmental significance of the proposed activity. For cases where the variation in the level of significance is much less, the decision on the classification of environmental impact of the proposed activity is likely be less. The significance of impact can be understood by naming low environmental impact, medium environment impact and high level of significance.

To understand such levels of significant impacts on environmental components like human health resources, air, water, flora, fauna and society, a right team of experts for the proposed project is very much essential and important for EIA study. Such team is also useful for the assessment of all significant impacts, direct, indirect as well as cumulative impacts associated with environmental resources. Generally, the EIA team consists of qualified and experienced professionals from the range of disciplines required for the EIA. The following is the broad list of experts that may be required depending on the environmental setting, project manager/team leader, geographer, geologist, hydrologist / hydro-geologist, hazard management specialist, chemist, marine biologist, botanist / floristic surveyor, ornithologist, herpetologist, lepidopterist, parks and protected areas specialist, urban / regional planner, socio-economist, demographer, impact assessment specialist the nature and category of the project. Fig. 8.1 shows the multidisciplinary EIA team.

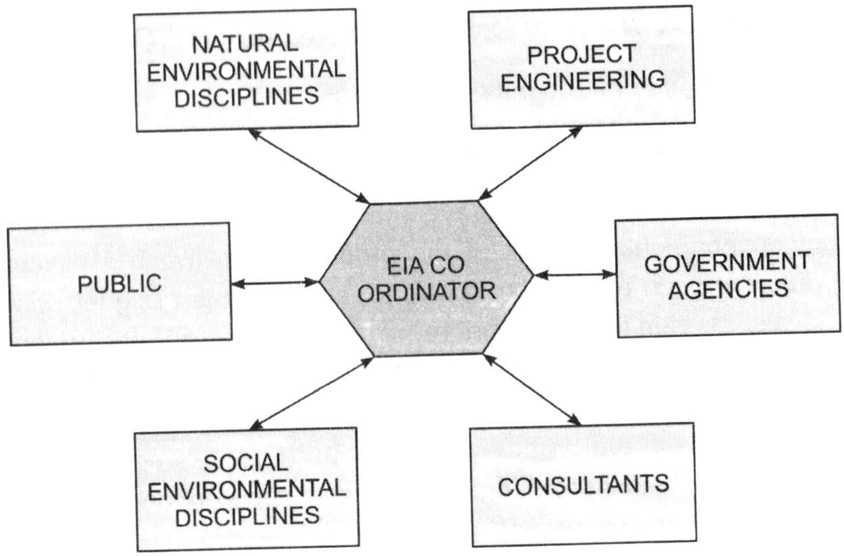

Fig. 8.1 Multidisciplinary EIA study team

8.6.1 Significance of Impacts

The significance of impacts first and proceeds to delineate the associated mitigations and measures. So the significance here reflects the "worst-case scenario" before mitigation is applied, and therefore provides an understanding of what may happen if mitigation fails or is not as effective as predicted. For establishing significance of different impacts, understanding the environmental system responses and interaction is essential. Hence, the impact interactions and pathways are to be understood and established first. Such an understanding will help in the assessment process to quantify the impact as accurately as possible. Complex interactions, particularly in the case of certain indirect or cumulative impacts, may give rise to non-linear responses which are often difficult to understand and therefore it might be difficult to assess their significance. It is hence understood that indirect or cumulative impacts are more complex than the direct impacts and most often currently the impact assessments are limited to direct impacts. In case mitigation measures are delineated before determining significance of the effect, the significance represents the residual effects,

However, the ultimate objective of an EIA is to achieve sustainable development. The development process shall invariably cause some residual impacts even after implementing an EMP effectively has a threshold for absorbing deterioration and a certain capacity for self-regeneration. These thresholds based on concept of carrying capacity are as follows:

- Waste emissions from a project should be within the assimilative capacity of the local environment to absorb without unacceptable degradation of its future waste absorptive capacity or other important services.
- Harvest rates of renewable resource inputs should be within the regenerative capacity of the natural system that generates them; depletion rates of nonrenewable inputs should be equal to the rate at which renewable substitutes are developed by human invention and investment.

The aim of this model is to curb over-consumption and unacceptable environmental degradation. But because of limitation in available scientific basis, this definition provides only general guidelines for determining the sustainable use of inputs and outputs. For establishing the level of significance to each identified impact, a three-stage analysis may be referred: First, an impact is qualified as being either negative or positive. Second, the nature of impacts such as direct, indirect, or cumulative is determined using the impact network and third, a scale is used to determine the severity of the effect; for example, an impact is of low, medium, or high significance.

It is not sufficient to simply state the significance of the effect. This determination must be justified, coherent and documented, notably by a determination methodology, which must be described in the methodology section of the report. There are many recognized methodologies to determine the significance of effects.

8.6.2 Criteria to determine the Significance of the Identified Impacts

The criteria can be determined by answering some questions regarding the factors affecting the significance. This will help the ETA stake-holders, in particular, practitioner to determine the significance of the identified impacts eventually. Typical examples of such factors on include the following:

- *Exceedance of a threshold:* Significance may increase if a threshold is exceeded. e.g.. Emissions of SO_2 and/or PMIO exceed the permissible threshold.
- *Effectiveness of Mitigation:* Significance may increase as the effectiveness of mitigation measures decreases. e.g., control technologies, which may not assure consistent compliance to the requirements.

- *Size of study area:* Significance may increase as the zone of effects increases e.g. High temperature discharge from a cooling tower in sea may impact the mangrove ecology at a distant location.
- *Incremental contribution of effects from action under review:* Significance may increase as the relative contribution of an action increases.
- *Relative contribution of effects of other actions:* Significance may decrease as the significance of nearby larger actions increase.
- *Relative rarity of species:* Significance may increase as a species becomes increasingly rare or threatened.
- *Significance of local effects:* Significance may increase as the significance of local effects is high.
- *Magnitude of change relative to natural background variability:* Significance may decrease if effects are within natural assimilative capacity or variability.
- *Creation of induced actions:* Significance may increase as a induced activities also highly significant and
- *Degree of existing disturbance:* Significance may increase if the surrounding environment is pristine:

For determining significance of impacts, it is important to remember that secondary and higher order effects can also occur as a result of a primary interaction between a project activity and the local environment. Where a primary effect is identified, the practitioner should always think if secondary or tertiary effects on other aspects of the environment could also arise.

The EIA should also consider the effects that could arise from the project due to induced developments, which take place as a consequence of the project. e.g. Population density and associated infrastructure and jobs for people attracted to the area by the project. It also requires consideration of cumulative effects that could arise from a combination of the effects due to other projects with those of other existing or planned developments in the surrounding area, so a qualitative checklist is suggested to test significance, in general. The checklists and matrices are beyond the scope of this book Reader may refer other books.

8.7 PREDICTION OF IMPACTS

The environmental impacts can be identified based on the activities in the proposed industry or plants or action or any plan. Based on the base line environmental status at the project site, the environmental components that may be affected (impacts) are identified. Both positive (beneficial) and negative impacts are to be considered for EIA study. Prediction of impacts is the most important component in the environmental impact assessment studies. Several scientific techniques and methodologies are available to predict impacts of developmental activities on physico, chemical, ecological and socio-economic environment. Such predictions are superimposed over the baseline data of environmental quality to derive the environmental scenario. The prediction of impacts helps to prepare the Environmental Management Plan (EMP) to be executed during the ongoing activities for the project to minimize the adverse impacts on environmental quality. The potential for environmental pollution during construction phase depends on aerial extent and its control of pollution is of considerable importance. The following factors require to control during the construction phase.

Site preparation: The clearance of site involves the movement of soil, leveling operations, control of dust nuisance created by excavation, transportation activities and facility of water sprinkling to reduce dust nuisance.

Sanitation: The construction site should be provided with sufficient and suitable toilet facilities for workers to allow proper standards of hygiene. These facilities should preferably be connected to a septic tank and maintained to ensure minimum environmental impact.

Noise: Though the noise effect on the nearest inhabitants due to construction activity will be negligible, it is advisable that onsite workers using high equipment adopt noise protection devices like ear muffs.

Construction equipment and Waste: It should be ensured that both gasoline and diesel powered construction vehicles are properly maintained to minimize smoke in the exhaust emissions. The vehicle maintenance area should be located in such a manner to prevent contamination of surface and ground water sources by accidental spillage of oil. Unauthorised dumping of waste oil should be prohibited. Wastes should be disposed off at an approved dumpsite.

Storage of Hazardous material: The hazardous materials like: petrol and diesel, LPG, painting materials if used for site preparation and construction activity should be stored as per international standards.

Site Security: Construction site is a potential hazardous environment. To ensure that the local inhabitants and stray cattle are not exposed to these hazards. The site should be secured by fencing and manned entry points.

8.7.1 Air Environment

"Air pollution" can be defined as the presence in the outdoor atmosphere of one or more contaminants (pollutants) in such quantities and of such duration as may be (or may tend to be) injurious to human, plant, or animal life, or to property (materials), or which may unreasonably interfere with the comfortable enjoyment of life or property, or the conduct of business. It should be stressed that the attention in this definition is on the outdoor, or ambient, air, as opposed to the indoor, or work environment, air. This chapter is focused upon air quality analysis with regard to the ambient atmosphere.

Air pollution can be caused by the presence of one or more contaminants. Examples of traditional contaminants include sulfur dioxide, oxides of nitrogen, carbon monoxide, hydrocarbons, ozone, oxidants, hydrogen sulfide, particulate matter, smoke, and haze. This list of air pollutants can be divided into two categories: gases and particulates. Gases, such as sulfur dioxide and oxides of nitrogen, exhibit diffusion properties and are normally formless fluids which can be changed to the liquid or solid state only by a combined effect of increased pressure and decreased temperature. Particulates represent any dispersed matter, solid or liquid, in which the individual aggregates are larger than single small molecules (about 0.0002 micrometers, μm, in diameter) but smaller than about 500 μm. (one micrometer is 10^{-4} cm.) Of recent attention is particulate matter which is equal to or less than 10 μm in size, with this size range of concern relative to potential human health effects. Additionally, particles in the atmosphere can have a lifetime ranging from only a few minutes to several months; larger particles settle more rapidly than smaller particles.

The focus is on "air toxics," or hazardous air pollutants. Air toxics are a class of compounds which may be present in the atmosphere and exhibit potentially toxic effects not only to humans but also to the overall ecosystem. In the 1990 Clean Air Act, the air toxics category includes 189 specific chemicals which may be of relevance in an air quality impact study (Quarles and Lewis, 1990), This group of compounds represents typical compounds of concern in the industrial air environment, with these compounds and associated relevant quality standards being adjusted for outdoor atmospheric conditions.

The above definition also mentions the quantity or concentration of the contaminant in the atmosphere, and its associated duration or period of occurrence. This is an important concept in that pollutants that are present at extremely low concentrations and for short time periods may be insignificant in terms of the planning and conduction of an impact study.

Additional air pollutants or atmospheric effects which have become of concern include photochemical smog, acid rain, and global warming. "Photochemical smog" refers to the formation of oxidizing constituents, such as ozone, in the atmosphere as a result of the photo-induced reaction of hydrocarbons (or volatile organic chemicals) and oxides of nitrogen. This phenomenon was first recognized in Los Angeles, California, in the period following World War II, and it has become a major air- pollution concern throughout the United States. "Acid rain" refers to atmospheric reactions which can lead to precipitation which exhibits a pH value less than the normal pH of rainfall (which is approximately 5.7 when carbon dioxide equilibrium is considered). Within recent years, in central Europe and in several Scandinavian countries, along with Canada and the northeastern United States, attention has been directed to potential environmental consequences of acid precipitation. Causative agents in acid-rain formation are typically associated with sulfur dioxide emissions and, possibly, nitrogen oxide emissions, along with gaseous hydrogen chloride. From a worldwide perspective, sulfur dioxide emissions are the dominant precursor of acid-rain formation.

Another issue of importance from a global perspective is the influence of air pollution on atmospheric heat balances and associated absorption or reflection of incoming solar radiation. As a result of increasing levels of carbon dioxide and other carbon-containing compounds in the atmosphere, there is a growing concern that the earth's surface has already started exhibiting increased temperature levels, and this, in turn, can have major implications in terms of shifting climatic conditions throughout the world.

Sources of Air Pollutants

Air pollutant sources can be categorized from several perspectives, including the type of source, their frequency of occurrence and spatial distribution, and the types of emissions. Characterization by source type can be delineated as arising from natural sources or from man-made sources. "Natural sources" include plant pollens, windblown dust, volcanic eruptions, and lightning-generated forest fires. "Man-made sources" can include transportation vehicles, industrial processes, power plants, construction activities, and military training activities.

Source characterization according to number and spatial distribution can include such categories as single or point sources (stationary), area or multiple sources (stationary or mobile), and line sources. "Point sources" are characteristic of pollutant emissions from industrial- process stacks, as well as fuel-combustion facility stacks. "Area sources" include vehicular traffic, fugitive-dust emissions from resource- material stockpiles or construction, or military training activities over large geographical areas. Figure 6.1 includes one delineation of source categories which can be used for analyzing air pollutant sources in a given geographical area.

Effects of Air Pollutants: Air pollution effects may also be divided into several categories, with such effects encompassing those that are health-related as well as those associated with damage to property or materials or which cause decreases in atmospheric aesthetic features. 'Examples of effects on human health include eye irritation, headaches, and aggravation of respiratory difficulties. Plants and crops have been subjected to the undesirable consequences of air pollution, including abnormal growth patterns, leaf discoloration or spotting, and death. Animals such as cattle have been subjected to undesirable consequences of atmospheric fluorides. Property and materials damages include property devaluation because of odors, deterioration of materials such as concrete statuary, and discoloration of painted surfaces on cars, buildings, and bridge structures.

The aesthetic effects include reductions in visibility, discoloration of air, photochemical smog-related traffic disruptions at airports, and the general nuisance aspects of odors and dust.

Air Quality Modeling: Air quality models are the primary tools for relating emissions to air quality impacts. Models, in turn, require acceptable input data for emissions, surface topography, meteorological parameters, receptor configurations, baseline air quality and initial and boundary conditions for each modeling scenario. Since the quality and reliability of model outputs can never be any better than the inputs, quality control of the input data is an important concern. The baseline air quality and emission behavior of the project have been characterized before structuring the air quality impact analysis. The baseline air quality may be characterized a representative background air quality. Prediction of impacts on ambient air environment has been carried out employing mathematical model based on a steady state Gaussian plume dispersion model designed for point sources for short term. For instance, industrial Source Complex [ISC3] 1993 dispersion model based on steady state Gaussian plume dispersion, designed for point sources for short term and developed by United States Environmental Protection Agency [USEPA] has been used for simulations from point sources. The model simulations deal with the major pollutant viz., PM, SO_2 and NO_x emitted from the industrial activity. The Gaussian Plume Model (ISCST3) is the short term area source model is based on a numerical integration over the area in the upwind and cross wind directions of Gaussian plume formula. This can be applied to the point, area, line or volume sources simultaneously and their resultant incremental concentration of the pollutant can be predicted. The meteorological data is very much needed in predicting air pollution caused by any proposed project / activity.

Meteorological data: Data recorded at the manual weather monitoring station on wind speed, direction, and temperature at one hour interval for the specified monitoring period. The conditions to be adopted in the process of predictions air pollution impacts are :

- For the prediction of rise in ground level concentrations of pollutants, the actual meteorological data recorded near the plant site during the study period is to be used.
- Predictions are carried out for 100% load with control equipments in operation.
- Predictions are carried out for Particulate matter, SO_2 and NOx emissions.
- No half life time of the pollutant is considered.

The stack height has 30 m for the 3.0 TPH & 2 TPH ensures good dispersion of pollutants in the atmosphere thereby having insignificant impacts on the ground level concentration. The stack height should meet the MoEF guidelines. To symmetrically address the air quality impacts of potential projects or activities, it is necessary to be familiar with basic information regarding air pollution. Accordingly, this section includes a definition of air pollution and information on the types and effects of specific air pollutants, along with summary information on air pollution sources, which are given in above paras.

A site–specific background air quality monitoring program was conducted for the existing project site during the month of November 2009. Background data was collected for SPM, RSPM, SO_2, NOx. The basic considerations for designing air quality surveillance programme include:

(i) Topography
(ii) Physical Features
(iii) Micro-meteorology of the region
(iv) Representation of regional background
(v) Proper representation of upwind and downwind directions
(vi) Ecologically Sensitive Locations within 5 km

Keeping the historical meteorological conditions, topography, physical features, sensitive locations and current and anticipated pollution loads in mind, and based on the reconnaissance Survey of the Project Area, a monitoring framework for air quality in the study area (10 km radius) is to be drawn.

Ambient Air Quality: An assessment of baseline air quality was undertaken to (a) establish the status of exposure of the major sensitive receptors, and (b) to, identify the major air pollution sources and their impacts on the area surrounding the project site. This assessment was accomplished by reviewing historical ambient air quality data examining sources of air emissions within 5 km radius of the proposed site (i.e. the impact zone), and by conducting a site-specific background-sampling program. In this manner, background data collected was expected to be representative of all meteorological conditions. The National Ambient Air Quality standards can be seen in the chapter 7.

The baseline studies for air environment covers reconnaissance, identification of specific air pollutants expected to have significant impacts from the proposed project and assessing their prevailing levels in ambient air at representative locations within the impact zone around the project site. The baseline status of air environment in this study has been assessed through reconnaissance in project area and a systematic air quality surveillance programme.

8.7.2 Water Environment

The major activities in the project area delineated around the proposed site (5 km radial distance) are residential activities. The study area is a flat terrain. Water sampling has to collect to determine the existing quality of water around the project area and also to assess the impact due to proposed project. Sampling should be done following the standard guidelines for physical, chemical and bacteriological parameters. Water sample from both surface as well as ground water sources were collected from locations in the vicinity of project site.

The ground water samples analysis results are compared with the IS : 10500 desirable and permissible limits, which is stipulated for water to be fit for drinking purpose with ground water as source. The desirable and permissible limits of various parameters as per the IS 10500 standards are to be analyzed. The chemical examination of the water samples reveal that all the locations pH, colour, odour, taste, Iron as Fe, Cadmium as Cd, Lead as Pb and Arsenic as (As) are within the desirable limits. Total Hardness for all the samples except at GW-4 is above the desirable limit of 300 mg/L, however below the permissible limits. The concentrations of Chlorides (as Cl) and Nitrates (as NO_3) for all the samples is within the desirable limits. The concentration level of Fluorides (as F) for all the samples is above the desirable limit 1.Omg/L and below or equal to permissible limits at all the locations except from samples collected from GW-4, where fluoride concentration is 1.61 mg/L. From the available data it may be considered that the ground water from the locations GW-1, GW-2, GW-3 and GW-5 may be used for drinking, if no other alternate source of water is available, whereas the ground water from the location GW-4 is unfit for drinking, requires suitable treatment to make it potable.

The surface water analysis results are compared with the tolerance limits for Inland surface water bodies (IS 2296-1982), Class — C, Drinking water source with conventional treatment followed by disinfection. From the results it can be seen the pH value and the concentrations of Total dissolved solids, Chlorides, Sulphates, Iron, Lead Arsenic, Zinc are all well within the standard limits, however the BOD levels at all three locations are either equal to or higher than the stipulated limits. Among the three samples collected, the sample collected from the location SW-2 is showing a highest BOD level of 5 mg/L. From the analytical results it is seen that the total coli forms are well within the standard limits. In general the surface water quality collected from all locations requires suitable treatment.

Surface water bodies like rivers, streams, canals, reservoirs, estuaries, coastal waters, lakes etc. play very important role in the sustainability of any ecosystem. It is very important to assess the impacts of any developmental activity on these surface water environments. Impacts on surface waters are usually caused by physical disturbances; i.e., the construction of banks, dams, and other natural or man-made drainage systems. The discharge of effluents and deposition of air pollutants into water also causes pollution. These activities and processes effects as follows: (a) changes in surface water hydrology, (b) changes in surface water quality, (c) changes in sediment behavior, (d) changes in salinity and (e) changes in aquatic ecology.

8.7.3 Noise Environment

Any unpleasant sound is classified as noise pollution. Sound possesses three definite properties: intensity, frequency and duration. Intensity is the loudness of a sound, or the pressure it exerts through the ear. It is measured in decibels (dB). In assessing noise, a special measure callod "dB (A)" indicates damage to hearing. The higher the dB (A) number, the greater is the risk of damage to hearing.

Loud noise may adversely affect people in many ways. For example, noise may interface with sleep, speech, communication and can cause annoyance and other physiological problems. Occupational noise exposure, is also the most common cause of Noise-Induced Hearing Loss

(NIHL), threatens the hearing of individuals exposed to noise pollution for longer periods of time, at a less intense level. For example, repeated exposure to noise pollution at a construction site can cause NIHL to construction workers, an effect that cannot be reversed.

Compressors, Boilers and DG set will be the major noise generating units in the plant. Out of these the generator will be functioning only at the time of power failure. There is no need for the workers to be near this unit continuously. However the workers in this area are always provided with ear muffs. The effects of the vibrations coming out of the base of the body over the surrounding civil structures are minimized by providing damping pads/resilient mounting at the time of installation. The nearest settlements are about 0.5 km from the industry and the Irikepalli village which is about 1.0 km from the industry, no impact of noise will be felt at this village. Under the general health check up scheme as per factory act, the workers will however be checked up for any Noise induced Hearing Loss (NIHL) by a trained ENT Doctor. Noise Attenuation Achieved with 5-10 Meter Thick Greenbelt of Pure Species Monoculture:

All the equipment in the plant would be designed to have a total noise level not exceeding 85-90 dB(A) as per the requirement of OSHA (Occupational Safety and Health Administration) standards. The noise levels in the work place environment will be monitored periodically a, d if necessary action will be taken in the form of regular maintenance schedule to reduce noise and vibration in generating sources.

During the construction vehicles carrying construction material and machinery & equipments at site are expected to generate noise. During the operational phase of the site, the following sources of noise pollution are expected (Table 8.4).

Table 8.4 Primary Effects of Noise Pollution from Construction & Operation Activities of the Site

Pollutant	Source	Primary Effects
Noise	Construction activities Transport Human activities	Annoyance Interference with communication NIHL Interference with sleep Aural pain, nausea & reduced muscular control Performance effects Effects on social behavior

A detailed survey on noise environment is to be carried in and around the mine site to study the hourly equivalent noise levels as per IS: 4594 – 1968. Spot noise levels are measured for 24 hours on hourly basis by using a high precision sound level meter within the study area.

8.7.4 Human Health and Society

Impacts on human health and society are measured in terms of socio-economic impacts indirectly. Socio-economic impacts are equally important as biophysical impacts. Any development activities in a region are weighed by positive and negative socio-economic parameters. This can be explained by a case study for easy understanding to our graduate students.

National Environmental Engineering Research Institute (NEERI, 1991) Nagpur, prepared a report titled – 'Report on Environmental Pollution caused by industrial estates in near by villages of Medak district in Andhra Pradesh', based on studies and observations of five scientists. The reports concentrated mainly on human health and mortality, and their economic loses are due to crop failures and death of domestic animals like cattle, sheep, cows etc. due to industrial pollution. Industrialization has resulted in heavy discharge of toxic chemical effluents to various water resources such as streams, rivers and tanks causing serious damages to water quality and contamination of ground water and soil, agricultural production depends upon the quality of irrigation water. Human health is one of the most important factors for economic development. A healthy workforce is essential to the development of economy. Pearce and Warford (1993) have stated that the most important and immediate consequences of environmental degradation in the developing world takes the form of damage to human health. As cited by Pearce and Warford (1993) it is confirmed that mortality and morbidity from water borne diseases due to water pollution have definite impact on mankind. Ultimately, it has serious negative impact on economic activities in the form of loss of working days, deaths of trained workers and expenditure on hospitalization etc..

Water pollution is compared with the abatement cost pollution control program in Thailand, by direct questioning method along with the available scientific findings. The relationship between the damage cost and water quality are examined. It was concluded that the treatment level had to be kept above 75% in order to maintain river quality above 4 ppm of dissolved Oxygen above which no damage cost are assumed. This is a serious threat to a sustainability of these economics, as many of the countries fall under the negative growth when these cost are incorporated into their Gross National Product (GNP). The exact estimation of damage to cost depends upon the reliability of data relating to environmental damage. It is confirmed that environmental deterioration damages the economies of both the rich and the poor countries.

Irrigation systems are most affected by industrial pollution in Medak district. There are as many as 15 irrigation tanks. All the 15 tanks have been considerably polluted. As irrigation has no alternative for water, farmers are forced to use the same polluted water. As a result, the productive lands are turning unproductive. It is estimated that about 920 acres of agricultural land is affected. In study area paddy is the major crop, which gives employment to a sizeable amount of agricultural workers. This indicates that the impact is in terms of loss of production and employment losses.

Water, Air and Soil pollution have affected the villages around in this region. The most strident problem is that of drinking water. People are depending on the municipal drinking water supply, but the meager municipal water supply force the villagers to depend on the contaminated water for the purposes other than drinking. Livestock is also forced to drink the polluted water and hence suffer from various diseases. Number of livestock deaths also reported in the villages, apart from declining milk yields and fertility rates.

The industrialization in this study area started during 1986. According to the villagers, they started realizing the polluting in the 1990s. Initially the villagers thought that the industrial effluent would be beneficial for agricultural production, because they thought that the chemical water would provide more fertility to land. Between 1995-96 the villagers witnessed a rapid change of color of tank water and at the same time they witnessed the death of fish. By 1997 it was observed that there were no fish left in the tank. In 1998-99 farmers got to know about the impact of polluted water on crops (Rice crop was observed with empty husk with out rice grain in it, the size of potato is reduced by 50%). The impact of pollution on health is conspicuous. Most of the villagers are suffering from water borne diseases. The problem further escalated when the buffalos, cattle and cows became victims of various diseases by drinking polluted water from tanks and drains. More over the villages has been incurring heavy losses by losing bullocks, cows and other valuable bovine population. By 1998-99 villagers realized that their groundwater has become polluted and undrinkable.

The study indicated that the economic condition of the villagers has been of late quite pathetic. The main reason for this is that living conditions which were of above average a decade back, has been brought down due to rapid deterioration of health, agriculture and substantial reduction of livestock due to polluted water.

The entire Digwal and Chilkapally villages have been suffering from various diseases arising out of air, water pollution. However, it was observed that most of the diseases are water borne such as skin infection, teeth corrosion, joint pains, loss of appetite defective vision, fever, abnormal pain, respiratory diseases and diarrhoea etc., apart from that general muscular weakness, irregular growth, chronic cold and cough in the middle aged and children are noticed in the village. Most of the communicable diseases are due to polluted water, polluted soil. Diseases such as gastroenteritis, typhoid, viral infections are caused by oral-faecal contamination due to drinking infected water or food. People in many areas drink surface water from canals, tanks and rivers directly. Bathing and cleaning of animal are done at the same place where drinking water is collected. This greatly contributes to high incidence of water borne diseases, air born diseases such as tuberculosis, acute respiratory infections. Malaria is rampant through out the year. This is caused by the bite of anophiline mosquitoes that breed in open drainages and stagnant water pools.

Social assessment refers to a broad range of processes and procedures for incorporating social dimensions into development projects. In some jurisdictions and agencies, the social assessment is conducted in conjunction with the environmental impact assessment (EIA). In others, it is conducted separately. In any case, the social assessment influences project design and the overall approval of the project. People are to be considered vulnerable groups - some of those will be benefited and some groups may be harmed by the project activity.

The social assessment aims to determine the social costs of the project. The benefits of a project will be distributed in an equitable manner. Social assessments are necessary to help ensure the project will accomplish its development goals i.e., poverty reduction; enhancement of the role of women in development; human resources development, including population planning; and avoiding or mitigating negative effects on vulnerable groups, and protecting these groups. By addressing the specific development goals in the assessment of development projects, developers, and governments, can ensure that project benefits are realized and negative social impacts are minimized.

Various methods and approaches have been developed to consider social dimensions, they are:

(a) social and gender analysis;

(b) indigenous peoples plans; and involuntary resettlement plants;

(c) cooperation with non-governmental organizations;

(d) use of participatory development processes;

(f) benefits monitoring and evaluation.

8.7.5 Flora and Fauna

Flora: The consumption of land and the consequent loss of natural habitat is inherent in road development. Where new roads intersect habitat, the area occupied by the road itself, borrow pits, and quarries is subtracted from the total habitat area available to flora and fauna.

When a road cuts through an ecosystem, the sum of the two parts created by the cut is less than the value of the initial whole, even when the habitat loss is ignored. Ecosystems are characterized by complex, interdependent relations between component species and their physical environment, and the integrity of the ecosystem relies on the maintenance of those interactions. Roads tend to fragment an area into weaker ecological sub-units, thus making the whole area more vulnerable to invasions and degradation.

Fauna: Most animal species tend to follow established patterns in their daily and seasonal movements. The areas, through which they travel on their way to and from feeding, and between their seasonal ranges, are known as corridors. When a road intersects or blocks a wildlife corridor, the result is either cessation of use of the corridor because animals are reluctant to cross the road, an increase in mortality because of collisions with vehicles, or a delay in migration which may result in the weakening or disappearance of an entire generation of the population. Some animals are attracted to roads for various reasons, including protection from predators, good food supplies, better travel conditions, etc.

8.8 ENVIRONMENTAL IMPACT STATEMENT (EIS)

An EIS is a detailed written statement which serves as an action-forcing device to ensure that the policies and goals defined in the EPA are infused into the ongoing programs and actions of the central government. It must provide full and fair discussion of significant environmental impacts and must inform decision makers and the public of the reasonable alternatives which would avoid or minimize adverse impacts or enhance the quality of the human environment.

If it is determined that a central action has a significant impact or impacts based on the EA, or if the agency has previously listed the action as one typically requiring an EIS, then an EIS should be prepared. The EIS normally include the following actions requiring 1. Feasibility reports for authorization and construction of major projects 2. Proposed changes in projects which increase size substantially or include additional purposes, and 3. Proposed major changes in the operation and/or maintenance of completed projects.

Three types of EISs may be pertinent (draft, final, or supplement to either draft or final); in addition to the earlier generic definition of an EIS, the following information from CEQ guidelines or regulations is germane. (Larry W. Canter 1996).

8.8.1 Draft EIS

The drat EIS is the document prepared by the lead agency proposing an action; it is circulated for review and comment to other concerned agencies, state and local agencies, and public and private interest groups. The agency must make every effort to disclose and discuss at appropriate intervals in the draft statement all major points of view on the environmental impacts of the alternatives, including the proposed action.

8.8.2 Final EIS

The final EIS is the draft EIS modified to include a discussion of problems and objections raised by the reviewers. The final statement must be on file with the EPA for at least a 30-day period prior to initiation of construction on the project.

8.8.3 Supplemental EIS

Lead agencies are to prepare supplements to either draft or final EISs if the agency makes substantial changes in the proposed action that are relevant to environmental concerns; or if there are significant new circumstances or information relevant to environmental concerns and bearing on the proposed action or its impacts. Lead agencies may also prepare supplements when the agency determines that the purposes of the act will be furthered by doing so.

8.9 ENVIRONMENTAL MANAGEMENT PLAN (EMP)

To ascertain the impact of the project on those applicable environmental parameters, an appropriate plan should be developed and described for constant monitoring. These are specially sensitive for the project under consideration. Usually these will include environmental resources within the industrial plant i.e., occupational health, and those in the region affected by plant establishment and operations.

The Environmental Management Plan would consist of all mitigation measures for each item wise activity to be undertaken during the construction, operation and the entire life cycle to minimize adverse environmental impacts as a result of the activities of the project. It would also delineate the environmental monitoring plan for compliance of various environmental regulations. It will state the steps to be taken in case of emergency such as accidents at the site including fire.

The Environmental Management Plan (EMP) is a site specific plan developed to ensure that the project is implemented in an environmental sustainable manner where all contractors and subcontractors, including consultants, understand the potential environmental risks arising from the proposed project and take appropriate actions to properly manage that risk. EMP also ensures that the project implementation is carried out in accordance with the design by taking appropriate mitigative actions to reduce adverse environmental impacts during its life cycle. The plan outlines existing and potential problems that may adversely impact the environment and recommends corrective measures where required. Also, the plan outlines the roles and responsibility of the key personnel and contractors who are charged with the responsibility to manage the proposed project. It is suggested that as part of the EMP, comprising of the site in-charge/coordinator, environmental group representative and project implementation team representative should form a monitoring committee. This committee's role would be to ensure proper, operation and management of the EMP including the regulatory compliance.

The key benefits of the EMP are that it provides the organization with means of managing its environmental performance thereby allowing it to contribute to improved environmental quality. The other benefits include cost control and improved relations with the stakeholders. EMP includes four major elements, namely, Commitment and policy, planning, implementation and measurement and evaluation.

The components of the environmental management plan, potential impacts arising out of the project and remediation measures are summarized below in Table 8.5 for the proposed aerospace industry.

8.10 CASE STUDY

In this section, an attempt is made to present a case study which demonstrates the EMP for a proposed aerospace and precision engineering Special Economic Zone (SEZ) at Adibalta village. Ibrahimpatnam Mandal, Ranga Reddy District, Andhra Pradesh. This EIA study was conducted by Environmental Protection, Training and Research

Institute (EPTRI), Hyderabad for Andhra Pradesh Industrial Infrastructure Corporation (APIIC).

The Aerospace SEZ would attract various companies engaged in the business of maintenance, repair and overhauling of the aircrafts. It would also house other ancillary companies engaged in ancillary manufacturing, supported by integrated infrastructure facilities such as ready to use factory sheds, office space, world class testing facilities, training institutions, business centre and other support facilities.

The proposed project location map is shown in Fig. 8.2. The unique concept of developing the aerospace industry specific SEZ would sustain the functions of growth and future expansion of the aerospace industry by providing the requisite infrastructure and facilities. There are six key activities to be undertaken by the units to set up in the proposed SEZ of aerospace industry. They are: engine overhauling, ancillary overhaul and repair, ancillary design and development, technical services, component overhaul and repair and aero-university.

The study on the existing environmental status in the project region was undertaken to generate the baseline data within the buffer zone around the project site on micro-meteorology, air quality, ground and surface water quality, noise levels, land environment including soil quality, land use pattern, biological environment and socio-economic status of the population, etc.

8.10.1 Identification of Impacts

The primary function of an environmental impact assessment study is to predict and quantify potential impacts, assess and evaluate the magnitude. The impact assessment forms the basis for development of Environmental Management Plan. Environmental impacts could be positive or negative, direct or indirect, local or regional and also reversible or irreversible. The main aim of this section is to provide some exposure to how an EMP can be developed for any purposed project. Hence the information on the baseline environmental conditions at the proposed project site for various parameters, various pollution loads and stressors that could impact the environment are not given in this section. But it discusses the incremental environmental impacts on the environmental parameters during the construction and operational phases of the project. The potential impacts during construction and operational phase of the project have been identified and presented in Table 8.5. This table also provides mitigation measures through EMP to be adopted for minimizing the impacts.

Table 8.5 Summary of potential impacts and remedial measures

S.No.	Environmental components	Potential impacts	Potential source of impact	Controls through EMP and design	Impact evaluation
1.	Groundwater quality	Groundwater contamination	*Construction Phase* • Waste water generated from temporary make shift Labor camp site.	Septic Tank	No significant impact as majority of laborers would be locally deployed.
			Operation Phase • Sewage treatment	• Domestic sewage will be discharged into local sewer lines.	No negative impact on ground water quality envisaged. Not significant.
2.	Groundwater quantity	Groundwater depletion	*Construction Phase* • Use of Ground Water for construction activity,	• Bore well water will be used as a standby.	No significant impact on ground water quantity envisaged.
			Operation Phase • Water from Hyderabad Metropolitan Water Supply and Sewerage Board (HMWS&SB) is the source for domestic water supply. Other water requirements will be met from the recycled water. Bore well water will be used only as a standby.	• Rain Water Harvesting Scheme. • Storm water collection for water harvesting, • Percolation well to be introduced in landscape plan. • Awareness Campaign for reduced water use by employees	No significant impact on ground water quantity envisaged. The ground water harvesting scheme will enhance the ground water level in the area.

Table 8.5 *Contd…*

S.No.	Environmental components	Potential impacts	Potential source of impact	Controls through EMP and design	Impact evaluation
3.	Surface water quality	Surface water contamination	*Construction Phase* • Surface runoff from site during construction activity,	• Silt traps and other measures such as, additional on-site diversion ditches will be constructed to control surface run-off during site development.	No on site impact envisaged as surface water is present beyond the project site.
			Operation Phase • Discharge of domestic • wastewater to surface water body/land.	Domestic water treatment • scheme developed to treat the wastewater so that it can be reused for landscaping and flushing.	No off site impact envisaged
4.	Air quality	Dust Emissions	*Construction Phase* • All heavy construction activities,	• Suitable control measures will be adopted as per a dust control plan.	Not significant because dust generation will be temporary and will settle fast due to dust suppression techniques adopted
		Emissions of SPM, SO_2, NC_x and CO	*Construction Phase* • Operation of construction equipment and vehicles during site development, • D.G set operation (back up)	• Rapid on site construction and improved maintenance of equipment.	Not significant

Table 8.5 *Contd...*

S.No.	Environmental components	Potential impacts	Potential source of impact	Controls through EMP and design	Impact evaluation
			Operation Phase • Power generation through DG Set Operations. • • Emissions from vehicular traffic in operations.	• Use of ultra low sulphur diesel if available, • Use of clean fuel if available. Providing footpaths within the site for the employees. Plantation to be provided with specific species to help reduce SPM levels.	Not significant as D.G. sets would be operated as power back-up during power failure i.e for 2-3 hours only.
5.	Noise environment	Noise emissions	*Construction Phase* • Operation of construction equipment and vehicles during site development. • DG set operation (back up).	• Use of well-maintained equipment fitted with silencers, • Providing noise shields near the heavy construction operations. • Construction activity will be limited to daytime hours only,	Residential areas are located at a distance of 1 .0 km from the site and hence no significant impact due to increased noise levels is expected from project activities. Also, there are no sensitive receptors located within the near vicinity of site.
			Operation Phase • Noise from vehicular movement, • Noise from DG. Sets operation.	• Plantation of specific species and development of silence zones for traffic movement, • Providing acoustic enclosures for diesel generator sets.	Not significant as nearest receptor would be at a distance of 100 m from the main road. Short-term exposure within permissible limit.

Table 8.5 *Contd...*

S.No.	Environmental components	Potential impacts	Potential source of impact	Controls through EMP and design	Impact evaluation
6.	Land environment	Soil contamination	*Construction Phase* • Disposal of construction debris, • Waste handling from D.G sets.	• Construction debris will be collected and suitably used on site as per construction waste management plan.	Impact will be local, as any waste generated will be reused for construc-tion activities. Not significant.
			Operation Phase • Dumping of municipal solid waste on land • Waste oil handling from DC sets.	• The solid waste management system to be done by an authorized agency.	Since solid waste is handled on the site, waste dumping is not going to be allowed. Negligible impact
7.	Biological environment (Flora and Fauna)	Displacement of Flora and Fauna on site	*Construction Phase* Site Development during construction	The site is a barren land with sparse vegetation	
			Operation Phase Increase of Green Cover,	• Suitable green belt will be developed as per landscaping plan in and around the site using local flora, which will enhance the ecology	Positive impact
8.	Socio-economic environment	Population displacement and loss of Income	*Construction Phase* • No displacement of population		No negative impact

Table 8.5 *Contd...*

S.No.	Environmental components	Potential impacts	Potential source of impact	Controls through EMP and design	Impact evaluation
			Operation Phase • Site operation.	• Project will provide employment opportunities for about 20,000 persons and to the local people in terms of service personnel during operations	Positive impact
9.	Traffic pattern	Increase of Vehicular traffic	*Construction Phase* • Heavy Vehicular movement during construction.	• Heavy Vehicular movement will be restricted to daytime only and adequate parking facility will be provided.	No negative impact
			Operation Phase • Traffic due to employees once the site is operational.	• Vehicular movement will be regulated inside the site with adequate roads and parking lots in the site.	Minor negative impact

8.10.2 EMP for Air Environment

Construction phase: Environmental impacts during construction phase will be mainly due to civil works such as site preparation, RCC foundation, construction, material and machinery transportation, fabrication and erection etc. The construction phase impacts are temporary and localized phenomena except the permanent change in local landscape and land use pattern at the project site. However, they require due consideration with importance during project execution and also wherever applicable detailed procedures shall be implemented to prevent/mitigate adverse impacts and occupational hazards. To mitigate1the impact of SPM (dust) during the construction phase of the proposed project, two measures are recommended for implementation, namely, a dust control plan; and procedural changes.

8.10.2.1 *Dust Control Plan*

The following table provides a dust control plan, specific to construction activities.

Table 8.6 Dust control plan

Fugitive Dust Source Category	Dust Control Actions
Earth-moving	For any earth-moving which is more than 30m from all property lines, conduct watering as necessary to prevent visible dust emissions from exceeding 100m in length in any direction.
Disturbed surface areas (except completed grading areas)	Apply dust suppression in a sufficient quantity and frequency to maintain a stabilized surface; 2b Areas, which cannot be stabilized, as evidenced by wind driven dust, must have an application of water at least twice per day to at least 80 percent of the unstabilized area.
Disturbed surface areas (completed grading areas)	Apply water to at least 80 percent of all inactive accessible disturbed surface areas on a daily basis when there is evidence of wind driven fugitive dust.
Inactive disturbed surface areas	Apply dust suppressants in sufficient quantity and frequency to maintain a stabilized surface; OR Utilize any combination of control actions 2c, 2d, or 3a such that, in total, they apply to all inactive disturbed surface areas.
Unpaved roads	Water all roads used for any vehicular traffic at least twice per day of active operations; OR Water all roads used for any vehicular traffic once daily and restrict vehicle speed to 15 mph.

Table 8.6 *Contd...*

Fugitive Dust Source Category	Dust Control Actions
Open storage piles	Apply water to at least 80 percent of the surface areas of all open storage piles on a daily basis when there is evidence of wind driven fugitive dust; OR Install a three-sided enclosure with walls with no more than 50 percent porosity that extends, at a minimum, to the top of the pile.
Track-out control	Downwash of trucks (especially tyres) prior to departure from site.

The most cost-effective dust suppressant is water, because a source of water tends to be readily available on a construction site. Water can be applied using water trucks, handheld sprays and automatic sprinkler systems. Further more, incoming loads could be covered to avoid loss of material in transport, especially if material is transported off-site.

8.10.2.2 Procedural Changes

Idling Time Reduction: Construction equipment is commonly left idling while the operators are on break or waiting for the completion of another task. Emissions from idling equipment tend to be high, since catalytic converters cool down, thus reducing the efficiency of hydrocarbon and carbon monoxide oxidation. Existing idling control technologies, which automatically shut the engine off after a preset time can reduce emissions, without intervention from the operators.

Improved Maintenance: Recognizing that significant emission reductions can be achieved through regular equipment maintenance, contractors will be asked to provide maintenance records for their fleet as part of the contract bid and at regular intervals throughout the life of the contract. A monetary incentive/disincentive provision will be established to encourage contractors to comply with regular maintenance requirements.

Reduction of On-site Construction Time: Rapid on-site construction would reduce the duration of traffic interference and therefore, reduce emissions from traffic delay.

Operation Phase: To mitigate the impact of pollutants from diesel generator sets and vehicular traffic during the operational phase of the site are recommended for implementation: Diesel generator set emission control measures, Vehicle emission controls and alternatives; and Green belt development.

8.10.2.3 *Diesel Generator Set Emission Control Measures*

The proposed stack heights of 78 m are adequate to keep the air pollutants well within the prescribed limits; hence no additional emission control measures have been suggested.

8.10.2.4 *Vehicle Emission Controls and Alternatives*

Use of Clean Fuel by vehicles on-site (if available): The petrol used should be low sulphur, clean fuel (if available), which will reduce emissions on-site and in areas where employees are traveling.

Footpaths, and Pedestrian Ways: Adequate footpaths and pedestrian ways should be provided at the site to encourage non-polluting methods of transportation.

Adequate Car Parking Space: Parking will be planned in the block which is used exclusively for multi level car parking (MLCP) block.

8.10.2.5 *Greenbelt Development*

Increasing vegetation in the form of greenbelt is one of the preferred methods to mitigate air pollution. Plants serve as a sink for pollutants, act as a barrier to break the wind speed as well allow the dust and other particulates to settle out there. It also helps to reduce the noise level to some extent. The following Table 8.7 indicates various species of the greenbelt that can be used to act as a barrier.

Table 8.7 Trees that Can be Used as Barrier for Air Pollution

Air Pollution Attenuation	Dust Absorbers
Ficus glomerata (Guler) Terminalia tomentosa (Asan) Acacia auriculiformis (Ear leaf Acacia) Polyalthia lon gifolia (Debdaru) Ficus benghalensis (Banyan) Man gifera indica (Aam) Nerium odorata (Kaner)	Azadirachta indica (Neem) Melia azaderach (Mahanocm) Butea monosperma (Palash) Cessia fistula (Amaltas) Bauhınia variegata (Kachnar) Terminalia arjuna (Arjun)

8.10.3 EMP for Noisc Environment

Construction Phase : During the construction phase, there would be a temporary increase in ambient noise levels due to construction) machinery operation and movement of construction vehicles. Though the industrial countries have specified limits for occupational noise exposure, the permissible noise exposure limit for industrial workers is primarily concerned with harmful effects of noise and its objective is to protect the hearing of working people.

Exposure to continuous and intermittent noise levels louder than 115 dB(A) should not be permitted. Following mitigation / management measures shall be adopted during construction period:

- For protection of construction workers, earplugs should be provided to those workers who will be working very close to noise generation source.
- Servicing of all construction vehicles and machinery shall be done regularly and during routine servicing operations, the effectiveness of exhaust silencers shall be checked and if found defective will be replaced. Vehicles hired for bringing construction materials at sight shall conform to the noise emission standards and shall be operated during non peak hours.
- Workers employed in high noise areas will be rotated. Earplugs/muffs, or other hearing protective wear will be provided to those working very close to the noise generating machinery.
- Smooth flow of traffic should be ensured on the internal road to avoid idling and honking of vehicles.
- Ambient Noise level monitoring shall be conducted at suitable locations at periodic intervals during construction phase to conform to the stipulated standards both during day and night time. Data shall be reviewed and analysed by the project manager for adhering to any strict measure.
- Noise levels will also be monitored at point sources for occupational noise exposure and ensuring health risk.

Operation Phase: To mitigate the impact of noise due to vehicular movement during the operational phase the greenbelt development is recommended for implementation. The following species can be used as in a greenbelt to serve as noise breakers: Tectona grandis (Teak), Butea monosperma (Palash), Leucana leucocephala (Subabual), Mangifera indica (Aam), and Dalbergia Sissoo (Shisham), Azadirachta indica (Neem), Gre villea robusta (Silver Oak), Erythrina indica, Terminalia arjuna (Arjun), Aistonia scholaris (Devil's tree), Pterospermum acerifolium (Muchkund)

8.10.4 EMP for Water Environment

Construction phase: To prevent degradation and to maintain the quality of the water source, adequate control measures have been proposed to check the surface run-off, as well as uncontrolled flow of water into any water body. Following management measures are suggested to protect the water quality during the construction phase.

- Avoid excavation during monsoon season
- Care should be taken to avoid soil erosion.
- Community toilets with temporary septic tanks shall be constructed on the site during construction phase to prevent wastewater from entering the water bodies.
- Any area with loose debris within the site shall be planted.
- To prevent surface and ground water contamination by oil/grease, leak proof containers should be used for storage and transportation of oil/grease. The floors of oil/grease handling area should be kept effectively impervious. Any wash off from the oil/grease handling area or workshop shall be drained through impervious drains, Clarifiers or oil/water separators shall be constructed and effluent should be treated appropriately before releasing it.
- Construction activities generate disturbed soil, concrete fines, fertilizer, oils and other wastes. On-site collection and settling of storm water, prohibition of equipment wash downs, and prevention of soil loss and toxic releases from the construction site are necessary to minimize water pollution.

Operation Phase: Sustainable ground water extraction plan be evolved considering the water demand of the project in its operation phase. Water conservation and development measures need to be taken including all possible potential methods for reuse and recycling of water. These could be in the form of, development of water sources, minimizing water consumption, prompting reuse of water after treatment and development of closed loop systems for different water streams.

8.10.4.1 *Water Source Development*

Water source development shall be practiced by installation of scientifically design Rainwater Harvesting System (RHS). The water can be used for supply of portable water after treatment or recharge of aquifers. Rainwater harvesting promotes self-sufficiency and fosters an appreciation for water as a resource.

8.10.4.2 *Minimizing Water Consumption*

Water consumption will be minimized by a combination of water saving devices and other domestic water conservation measures. Furthermore, to ensure ongoing water conservation, an awareness programme will be introduced for the workers. The following section discusses the specific measures, which shall be implemented:

8.10.4.3 *Domestic and Commercial Usage*

- Use of water efficient plumbing fixtures (ultra flow toilets and urinals, low flow sinks, water efficient dishwashers and washing machines). Water efficient plumbing fixtures use less water with no marked reduction in quality and service.
- Leak detection and repair techniques.
- Awareness campaign to disseminate knowledge on strategies and technologies that can be used for water conservation.

8.10.4.4 *Horticulture*

- Drip irrigation system can be used for the lawns and other green area. Drip irrigation can save between 15-40% of the water use, compared with other watering techniques.
- Plants with similar water requirements shall be grouped on common zones to match precipitation heads and emitters.
- Use of low-volume, low-angle sprinklers for lawn areas.
- Select controllers with adjustable watering schedules and moisture sensors to account for seasonal variations, and calibrate them during commissioning.

8.10.4.5 *Storm Water Management*

As discussed earlier, most of the storm water produced on site will be harvested for ground water recharge, thus proper management of this resource is a must to ensure that it is free of contamination. Contamination of storm water is possible from the sources like: Diesel and oil spills in the Diesel power generator and fuel storage area, Oil spills and leaks in vehicle parking lots, Silt from soil erosion in gardens and Spillage of sludge from sludge drying area of sewage treatment plant.

A Storm Water Management Plan will be developed which will consider the above sources. The entire terrace rainwater and the surface storm water from the project site shall be collected through a RCC/Brick rectangular covered drain and recharged to the underground water table through a rain water harvesting system. The excess water after saturation of rain harvesting shall be discharged to the existing storm water drain outside the building premises. The RCC box drains with catch basins shall be provided along the periphery of the building and in open areas to collect rainwater.

The run-off from landscaped and roof top (hardscape) would be used for recharging the deep strata from which groundwater is to be extracted in the proposed site. The design of pits would be done taking into consideration retention time for the run-off water in the harvesting Pit.

The plan will incorporate best management practices, which will include the following:

- Regular inspection and cleaning of storm drains.
- Clarifiers or oil/water separators shall be installed in all the parking areas.
- Oil/water separators installed for parking areas and garages will be sized according to peak flow guidelines. Both clarifiers and oil/water separators must be periodically pumped in order to keep discharges within limits.
- Water conservation shall be practiced to the extent possible by use of reclaimed water for all non-potable application like gardening, watering lawns and floor washings etc.

8.10.5 EMP for Land Environment

Construction Phase: As discussed earlier, waste generated from construction activity includes construction debris, biomass from land clearing activities, waste from the temporary makeshift labour settlements, and hazardous waste. The following section discusses the management of each type of waste. Besides waste generation management of topsoil is an important area for which management measures are required.

8.10.5.1 *Construction Debris*

Construction debris is bulky and heavy and re-utilization and recycling is an important strategy for management of such waste. As concrete and masonry constitute the majority of waste generated, recycling of this waste by conversion to aggregate can offer benefits of reduced landfill space and reduced extraction of raw material for new construction activity. This is particularly applicable to the proposed project site as the construction is to be completed in a phased manner. Recycled aggregate will be used for filler application, and as a sub-base for road construction within the proposed SEZ. Mixed debris with high gypsum, plaster, shall not be used as fill, as they are highly susceptible to contamination, and will be given to recyclers.

Metal scrap from structural steel, piping, concrete reinforcement and sheet metal work shall be removed from the site by construction contractors. A significant portion of wood scrap can be reused on site. Recyclable wastes such as plastics, glass fiber insulation, roofing etc shall be sold to recyclers.

8.10.5.2 *Hazardous Waste*

Construction sites are sources of many toxic substances, such as paints, solvents, wood preservatives, pesticides, adhesives and sealants. Hazardous waste generated during construction phase shall be stored in sealed containers, labeled, and disposed of as required by the Hazardous Wastes Management and Handling Act Amendment Rules (MOEF, 2003). Some management practices to be developed are;

- Herbicides and pesticide will not be over applied and not applied prior to rain;
- Paintbrushes and equipment for water and oil based paints shall be cleaned within a contained area and shall not be allowed to contaminate site soils, watercourses, or drainage systems.
- Provide adequate hazardous waste storage facilities, hazardous waste collection containers are conveniently located, and designate hazardous waste storage areas are away from storm drains or watercourses.
- Segregate potentially hazardous waste from non-hazardous construction site debris.
- Clearly label all hazardous waste containers with the waste being stored and the date of generation.
- Educate employees and subcontractors on hazardous waste storage and disposal procedures.
- Instruct employees and subcontractors in identification of hazardous and solid waste.

Even with careful management, some of these substances are released into air, soil and water, and many are hazardous to workers. For these reasons, the best choice is to avoid their use as much as possible by using low-toxicity substitutes and low VOC (volatile organic compound) materials.

8.10.5.3 *Waste from temporary Labour settlements*

Waste generated from temporary labour settlements will be minimal as most of the labour deployed will be locals. The waste generated would mainly comprise of household domestic waste, which shall taken care by the contractors for collection and disposal at the nearest waste collection site.

8.10.5.4 *Top Soil Management*

To minimize disruption of soil and for conservation of topsoil, the contractor shall take the topsoil out separately and stockpile it. After the construction activity is over, topsoil shall be utilized for landscaping activity. Other measures, which would be followed to prevent soil erosion and contamination include

- Maximize use of organic fertilizer for landscaping and green belt development.
- To prevent soil contamination by oil/grease, leak proof containers should be used for storage and transportation of oil/grease and wash off from the oil/grease handling area shall be drained through impervious drains and treated appropriately before disposal.
- Removal of as little vegetation as possible during the development, and revegetation of bare areas after the project.
- Working in the area at a point of time (phase wise construction).
- Construction of erosion prevention troughs/berms.

Operational phase: The philosophy of solid waste management within the project during operation phase will be to encourage the four R's of waste i.e. waste reduction, reuse, recycling, and recovery (materials & energy). Regular public awareness meetings will be conducted to involve the workers and the waste management agency for proper segregation, storage, recycling options and techniques. The Environmental Management Plan for the solid waste focuses on three major components during the life cycle of the waste management system i.e. collection and transportation, treatment or disposal and closure and post closure care of treatment/disposal facility.

8.10.5.5 *Collection and Transportation*

- During the collection stage, the biodegradable and non-recyclable/non biodegradable waste will be stored separately. Thereafter the authorized agency will handle the waste management and disposal aspects.
- To minimize littering and odours, waste will be stored in well-designed containers/bins that will be located at strategic locations to minimize disturbance in traffic flow.
- Care should be taken so that the collection vehicles are well maintained and minimize noise and emissions. During transporting of the waste, it will be covered to avoid littering.

8.10.6 EMP for Biological Environment

In the present study, the impact of the project on flora and fauna needs to be addressed taking into account both in and around project area. The developmental work will certainly affect the ecological balance of flora and fauna present in and around the project site. There is no rare and threatened flora and fauna in the project site; however the local flora and fauna may also be conserved. It is important that adequate protection measures need to be followed.

- Wherever possible do not disturb the flora and fauna
- If any flora is disturbed, uproot and plant in open land near by
- Do not kill any fauna, as it will alter the biodiversity
- A green belt should be developed with trees and shrubs to mitigate dust pollution from the project area. The green belt will act as a barrier to trap the suspended dust particle and support the air pollution control.
- Regeneration of grassland area and soil conservation measure.
- If the flora is conserved, the fauna will not get disturbed.

8.10.7 Human Health and Safety Management Plan

The objective is to ensure that health and safety of on-site personnel is proactively managed during the construction stage of the project. Below are given the proposed project related Human Health and Safety environmental concerns and its management.

- The primary concern on potential health risks for the construction workers and other employees on site during construction are associated with drinking water quality. The project would ensure safe potable water supply to the workers on- site.
- Adequate space needs to be provided for construction of temporary sheds for construction workers to avoid unhygienic conditions.
- Construction site will be provided with a readily available first aid kit including an adequate supply of sterilized dressing materials and appliances. Suitable transport to take injured or sick person to the nearest hospital will be immediately provided.
- The project will ensure the safe working of all workers. Each construction worker will be provided with safety gadgets and made to

wear during the construction work. This will include protective footwear, helmets, and gloves to all workers employed for the work on mixing, cement, lime mortars, concrete etc.; the welder's protective eye-shields to workers who are engaged in welding works; earplugs to workers exposed to loud noise; safety belts to the labours working at higher platforms; and masks to avoid dust.

- The project will strictly follow the statutory child labour act. The project will also ensure that no paint containing lead or lead products is used except in the form of paste or readymade paint. Facemasks will be provided for use of the workers when paint is applied in the form of spray. Adequately safety measures will be ensured for workers during handling of materials at site. The project will comply with all regulations regarding safe scaffolding, ladders, working platforms, gangway, stairwells, excavations and safe means of entry and exit.
- The project will take adequate precautions to prevent danger from electrical equipments. No material will be so stacked or placed so as to cause danger or
- inconvenience to pay any person or the public. All necessary fencing and lights will be provided to protect the public. All machines to be used in the construction will conform to the relevant India Standard Codes, will be free from patent defect, will be kept in the good working order, will be inspected and properly maintained as per IS provision.
- Work spots will be maintained clean, and provided with optimum lighting.

8.10.8 Environmental Management System and Monitoring Plan

For the effective and consistent functioning of the proposed site, an Environmental Management System (EMS) should be established at the site. The EMS should include:

- Environmental management cell
- Environmental Monitoring
- Personnel Training
- Regular Environmental Audits and Corrective Action
- Documentation — Standard operating procedures Environmental Management plans and other records

8.10.8.1 *Environmental Management Cell*

Apart from having an Environmental Management Plan, it is also necessary to have a permanent organizational set up charged with the task of ensuring its effective implementation of mitigation measures and to conduct environmental monitoring. The major duties and responsibilities of Environmental Management Cell shall be as given below:

- To implement the environmental management plan,
- To assure regulatory compliance with all relevant rules and regulations,
- To ensure regular operation and maintenance of pollution control devices,
- To minimize environmental impacts of operations as by strict adherence to the EMP,
- To initiate environmental monitoring as per approved schedule.
- Review and interpretation of monitored results and corrective measures in case monitored results are above the specified limit.
- Maintain documentation of good environmental practices and applicable environmental laws as ready reference.
- Maintain environmental related records.
- Coordination with regulatory agencies, external consultants, monitoring laboratories.
- Maintain of log of common complaints and the action taken

8.10.8.2 *Environmental Monitoring*

The purpose of environmental monitoring is to evaluate the effectiveness of implementation of Environmental Management Plan (EMP) by periodically monitoring the important environmental parameters within the impact area, so that any adverse affects are detected and timely action can be taken.

In consultation with the Andhra Pradesh State Pollution Control Board (APPCB), the project proponents will monitor ambient air quality, noise levels, groundwater quality and quantity, soil quality and solid wastes in accordance with an approved monitoring schedule. The monitoring protocol and location selection will have to done carefully. The monitoring sampling program should be discussed and approved by APPCB. A suggested monitoring protocol, based on the predicted impacts, is given in Table 8.8.

Table 8.8 Suggested Post Monitoring Programme

S.No	Type	Locations	Parameters	Period and Frequency
1.	Ambient Air Quality	Project Site	Criteria Pollutants: SO_2, NO_x, SPM, RSPM	24-hr average samples every quarter during construction and annual during operation phase.
2.	Stack emission monitoring	Stack of operating DG set.	SO_2, NO_x, SPM	Annual
3.	Groundwater (Potable testing)	Project Site	Drinking water parameters as per IS 10500.	Annual
4	Surface water	Nearest surface water body	Drinking water parameters as per IS 10500.	Annual
5.	Ambient Noise	Project site	dB (A) levels	Hourly Day and Night time Leq levels every quarter during construction phase and annual during operation phase.
6.	Soil quality	Project site	Organic matter, C, H, N, Alkalinity, Acidity, heavy metals and trace metal. Alkalinity, Acidity.	Annual
7.	Waste Characterization	Project Site	Physical & chemical composition	Annual

8.10.8.3 *Awareness and Training*

Training and human resource development is an important link to achieve sustainable operation of the facility and environmental management. For successful functioning of the project, relevant EMPs should be communicated to the human resources. Relevant personnel at site must be trained for the following:

- Collection, transport, treatment and disposal solid and hazardous waste.
- Techniques for waste minimization, water conservation & energy conservation.
- Applicable environmental, health and safety regulations and compliance requirements for the same.
- Functioning of the environmental management System including environmental monitoring, reporting & documentation needs.

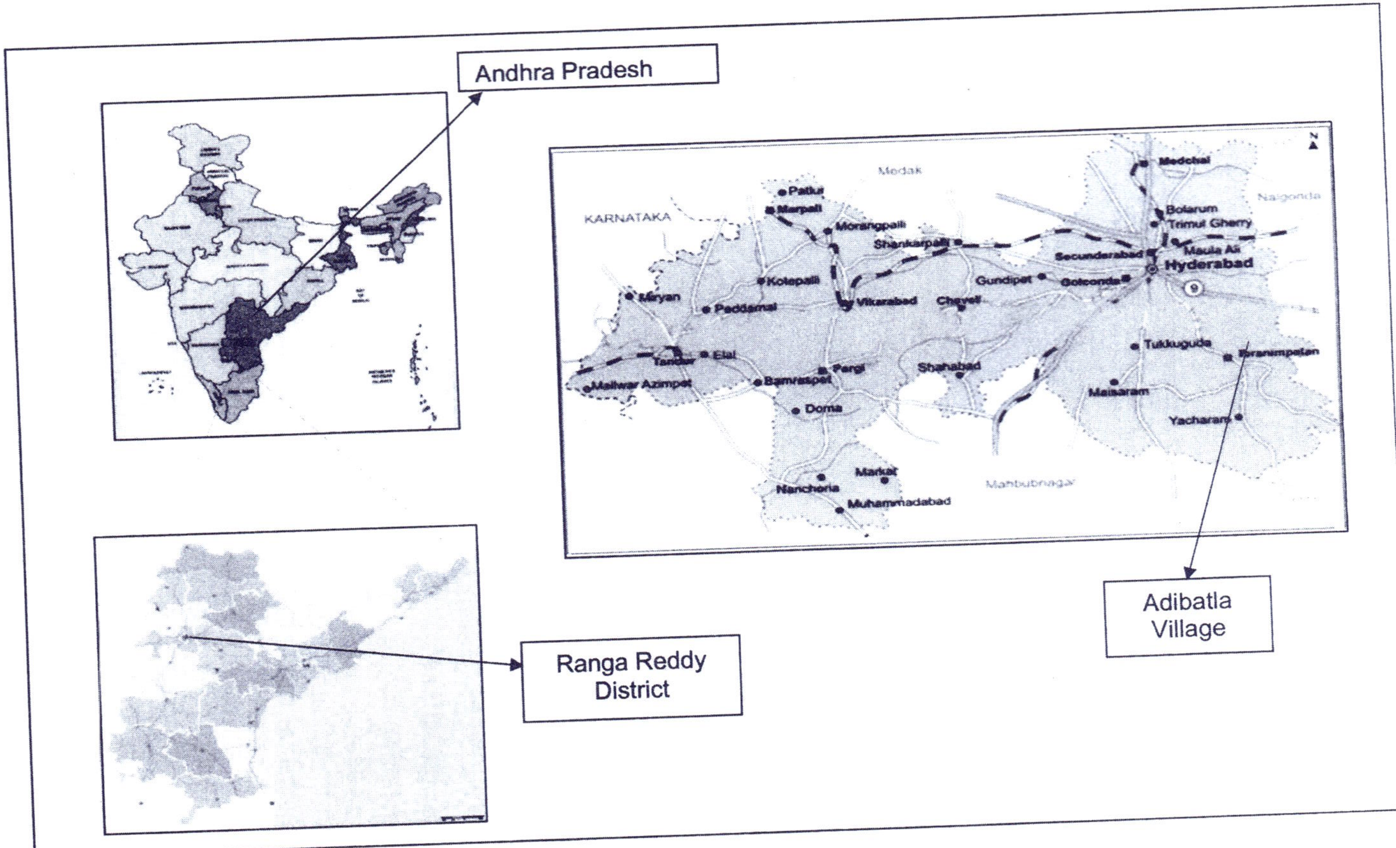

Fig. 8.2 Location map of the proposed project site

REVIEW QUESTIONS

1. Define EIA and explain EIA types.
2. Write briefly about scope of a comprehensive EIA.
3. Explain legal and educational roles of EIA.
4. List the tools of preventive environmental management.
5. Write an essay on environmental impacts.
6. How do you group all the impacts for systematic study.
7. Explain environmental inventory.
8. Explain Environmental Impact Statement.
9. Explain impacts on Environmental components.
10. Discuss the criteria to determine the significance of the identified impacts.
11. Write an essay on prediction of impacts on (i) Air environment (ii) Human health and society.
12. Write on essay on prediction of impacts on fauna and flora and water environment.

CHAPTER 9

TECHNOLOGICAL SOLUTIONS FOR POLLUTION CONTROL

9.1 GENERAL

An industrial plant sends a steady stream of exhaust gases into the air, and the wind carries the Pollutants away. If it didn't life in the vicinity of the plant would be nasty. But where is "away"? Is it the next mandal, district, next state, next country? The upper atmosphere. In reality, there is no "away" for the pollution of land, water and air resulting from the actinides of the billion of people inhabiting earth. Therefore, prevention of pollution or pollution is only the way through the existing and cost effective and executable technologies in the developing countries like India. Two basic strategies to bring the pollution under control are (i) reduce of remove the sources of pollution and (ii) treat and remove the pollution before it is released or convert them to harmless forms.

In this chapter, we will focus on some of the methodologies/technologies to prevent pollution and to control of water, air and soil resources.

9.2 SEWAGE TREATMENT PLANT TECHNOLOGY

The various treatment technologies used to treat wastewater at the sewage treatment plants of Delhi Jal Board (DJB) are discussed here. Although, there are 17 Sewage Treatment Plants under DJB, the technologies adopted at these plants can be categorized into 5 main type's i.e.

1. Conventional Activated Sludge Process
2. High rate Aeration with Biofiltration
3. Extended Aeration Process
4. Physico-Chemical Treatment with Biofiltration
5. Trickling Filters

The following Table 9.1 provides a list of the technologies along with the treatment plants using that technology:

Table 9.1 List of technologies with the plants being run on it.

S.NO.	TECHNOLOGY	SEWAGE TREATMENT PLANTS
1.	Conventional Activated Sludge Process	A. Rihala Phase-I B. Okhla C. Kondli D. Pappankalan E. Nilothi F. Narela G. Rohini H. Yamuna Vihar Phase-I I. Yamuna Vihar Phase-II J. Keshopur Phase-III K. Coronation Pillar Phase-II L. Coronation Pillar Phase-I
2.	High rate Aeration with Biofiltration	A. Rithala Phase-II
3.	Extended Aeration Process	A. Najafgarh B. Vasant Kunj C. Keshopur Phase-II D. Mehrauli E. Ghitorni F. Timarpur
4.	Physico-Chemical Treatment with Biofiltration	A. Sen Nursing Home B. Delhi Gate Nallah
5.	Trickling Filters	A. Coronation Pillar-I

9.2.1 Alternative Technologies to Sewage Treatment

Onsite Sanitation Options: The Conventional off-site excreta disposal and treatment method of water borne sewerage is an expensive option which is not affordable by low income communities and rural people. This is why a number of other onsite sanitation technologies were developed, practiced and popularized in due course of time. The most important among the two pit pour flush toilet technology developed and popularized by *Sulabh International Social Service Organisation.*

Some important on-site sanitation techniques are: septic tank, two pit pour flush toilets, trench latrine, bore hole latrine, dug well latrine, aqua privy and ventilated improved pit latrine (VIP).

Sulabh International, an NGO working in the field of sanitation from the last 33 years has also developed a few technologies to treat liquid waste/ sewage, besides the conventional techniques in sewage treatment plants. A brief review of these is presented below.

9.2.2 Sulabh Effluent Treatment (Set) System

Sulabh is maintaining over 5500 public toilet complexes spread all over the country, out of which 104 are linked with biogas plants. Therefore it was an important task for the organisation to make the effluent free from odour, colour and pathogens, so that it could be safely used for agricultural purposes. After a series of experiments, the organisation developed a new and convenient technology by which effluent of human excreta based biogas plant turns into a colourless, odourless and pathogen-free liquid manure. The technology is based on filtration of effluent through activated charcoal followed by ultraviolet (UV) rays. The filtration unit makes it colourless, odourless and free from organic particles and UV eliminates bacteria (Fig. 9.1). It reduces BOD, COD of the waste water drastically. Since the wastewater is from human wastes , its BOD (Biochemical Oxygen Demand) is around 200 mg/l, that comes down to <10 mg/l after treatment – safe for aquaculture, agriculture purposes, gardening or discharge into any water body. It can also be used for floor cleaning of public toilets in drought prone areas. Biogas from human excreta is being used for different purposes e.g. cooking, lighting, electricity generation and body warming. Besides, the liquid effluent can be used as fertilizer, as it contains good percentage of nitrogen, potassium and phosphate. The advantages of decentralised system of biogas plant with SET technology are:

- Costs on collection of sewage and operation and maintenance of the system are very slow.
- No manual handling of human excreta is required.
- Aesthetically and socially accepted.
- Biogas is used for different purposes.
- Treated effluent is safe to reuse as agriculture, gardening, or discharged into any water body.
- In draught prone areas treated effluent can be used for cleaning of floor of public toilets.

- If discharged into sewer, pollution load on STP will be much lowered. Thus, the decentralized system of sewage treatment through biogas technology is effective to minimize financial burden to combat the pollution.

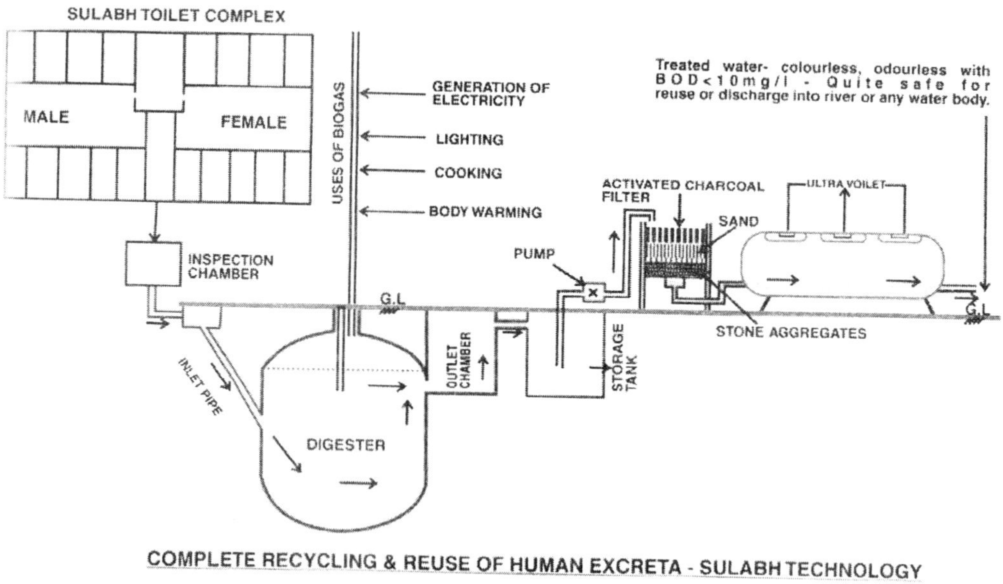

Fig. 9.1 Complete recycling and reuse of human excreta: Sulabh technology

9.3 DESIGN CRITERIA OF STP

The design of a sewage treatment work will be dependent on the quality and quantity of the waste to be treated. The following are some of the important characteristics of domestic sewage:

Organic Matter : Organic matter is the most important polluting constituent of sewage in respect of its effects on receiving water bodies. It is mainly composed of proteins, carbohydrates and fats. Organic matter is commonly measured in terms of BOD and COD. If untreated sewage is discharged into natural water bodies, biological stabilization of organic matter leads to depletion of oxygen in water bodies.

Nitrogen & Phosphorus: Nitrogen and phosphorus are also very important polluting constituents of sewage because of their role in algal growth and eutrophication of water bodies. Nitrogen is present in fresh domestic sewage in the form of proteinaceous matter urea (i.e. organic nitrogen). Its decomposition by bacteria readily changes it into ammonia. In aerobic environments ammonia nitrogen is oxidized into nitrites and nitrates. Nitrates can be used by algae to form plant proteins. Nitrogen is commonly measured as TKN (organic + ammonical) as sewage characteristics. Nitrate and nitrite forms of nitrogen are also measured when quality of receiving/affected water (streams, underground water) is monitored.

Phosphorus is usually present in orthophosphate, polyphosphate and organic phosphate forms. Organically bound phosphorus is of little importance in domestic sewage whereas polyphosphate forms undergo hydrolysis to revert into the orthophosphate forms, although this conversion is quite slow.

Suspended Solids : Suspended solids represent that fraction of total solids in any wastewater that can be settled gravitationally. Suspended solids can further be classified into organic (volatile) and inorganic (fixed) fractions. Organic matter is present in the form of either settleable form or non-settleable (dissolved or colloidal) form. If the organic fraction of suspended solids present in sewage is discharged untreated into streams, it leads to sludge deposits and subsequently to anaerobic conditions.

Dissolved Oxygen : Dissolved oxygen, as such, does not have any significance as a sewage characteristics. However, it is the most important pollution assessment parameter of the receiving water bodies. Stabilization of organic matter, when discharged untreated or partially treated in receiving waters, leads to depletion of their dissolved oxygen. Nutrients (nitrogen and phosphorus) addition due to discharge of untreated or treated sewage may lead to algal growth in streams. During day time, algae undergo photosynthesis process and the oxygen released by this process is much more than their respiration requirements resulting in a net addition of dissolved oxygen to water. However, during night time photosynthesis process is stopped whereas respiration requirement continues. This leads to depletion of dissolved oxygen in waters. Thus, it is observed that all the polluting constituents of sewage explained above have their direct or indirect effect on dissolved oxygen of receiving waters.

Bacterial Parameter (Fecal Coli form): Although organic matter, in dissolved as well as suspended form, is the most important parameter of sewage as far as ecology of receiving water bodies is concerned, Bacterial parameters, such as Fecal Coli form (FC), which serve as indicators of fecal pollution are also very important when human health is the prime concern. Sewage is discharged either into a water body, which is used for various purposes such as source of drinking water supply and bathing or discharged on land for irrigation, where human beings come in contact with it. Population consuming water from such sources which receive sewage discharges and persons involved in agricultural activities where sewage is applied become vulnerable to infection from pathogenic organisms (mainly bacteria and viruses) which are discharged by human beings who are infected with disease or who are carriers of a particular disease. Thus, to check quality of receiving waters for various uses and to assess acceptability of degree of treatment given to sewage, assessment of bacterial quality also becomes important. Because specific identification of pathogenic bacteria is extremely difficult, the coli form group of organisms is used as an indicator of the presence in wastewater of pathogenic organisms. Coli form bacteria are found in intestinal tract of human beings. Each person discharges about 100 to 400 billion coli form bacteria per day. Presence of coli form organisms is taken as an indication of presence of pathogenic organisms and absence of coli form organism is taken as an indication that water is free from disease producing organisms.

A detailed study on the sewage from various sources of study area is to be carried out and hence average characteristics of raw sewage can be obtained to design the treatment system (Table 9.2).

Table 9.2 Characteristics of waste water

Raw Sewage Characteristics

Parameter	Unit	Value for Design
pH	-	6.5 to 8.0
TSS	mg/lt.	200 to 250
COD	mg /lt.	300 to 450
BOD $_{3\ \text{days at}\ 27^{0}\text{C}}$	mg /lt.	150 to 300
Oil & Grease	mg/lt.	20 to 50

Treated Sewage Characteristics

Parameter	Unit	Value
pH	--	7.0 to 8.0
Suspended Solids	mg/lt.	< 30
COD	mg /lt.	< 150
BOD $_{3\ \text{days at}\ 27^{0}\text{C}}$	mg /lt.	< 30
Oil & Grease	mg /lt.	< 10

9.4 TREATMENT SCHEMES OVERVIEW

There is a large variety of treatment techniques designed to remove pollutants from wastewater. The objective of wastewater treatment is to separate wastes from water. In one sense, all wastewater treatment processes can be considered separation processes. There are physical, chemical and biological separation processes. Sedimentation and screening are examples of physical processes. Coagulation, ion exchange and pH adjustment are typical chemical processes, while various forms of biological digestion belong to the category of biological processes. In the biological processes living organisms, while in the physical and chemical processes physical and chemical properties are utilized for waste separation metabolizes organic wastes.

Treatment of sewage is accomplished by adopting various treatment schemes, each incorporating one or several different treatment units such as Screens, Grit chambers, Plain Sedimentation, Chemical Precipitation, Trickling Filter, Activated Sludge, Anaerobic digestion, Up flow Anaerobic Sludge Blanket (UASB) reactor, Waste Stabilization Pond and Maturation Pond. It has been found that low capital and low operational cost sewage treatment method such as Waste Stabilization Ponds OP or WSP technology and low operational cost sewage treatment method such as UASB technology are quite effective in BOD removal as well as Fecal Coli form (FC) removal. Overall efficiency of STPs based on these low cost technologies in terms of BOD and FC removal can be further improved if suspended solids (SS) are controlled by improvement in final outlet structures. These technologies are best suited for larger quantities.

Instead of conventional treatment systems latest advanced treatment schemes are available like Submerged Aerobic Fixed Film (SAFF) reactors, Fluidized Aerobic Bioreactor (FAB) and Membrane Bioreactors (MBR) . Advantage of these systems are they are easy to operate & maintain, efficient results, less foot print, skid mounted type packaged units.

9.4.1 Upflow Anaerobic Sludge Blanket (UASB)

Among the high rate reactors for wastewater treatment, the UASB process has gained popularity in earlier years all over the world. Several distilleries in the country have also adopted UASB treatment system because of its advantage over the conventional treatment. In the last 20 years, a large number of UASB units have been built in the world for treating high BOD industrial wastes (distilleries sugar, milk etc.). Since 1982, their use has been extended to include typical municipal sewage which has a relatively low BOD of only 200-300 mg/l. The advantages of UASB are:

- The hydraulic retention time is only 8-10 hrs.
- No prior sedimentation is required.
- The anaerobic unit does not need to be filled with any stones or other media, the up flowing sewage itself forms millions of small "granules" which are held in suspension and thus provide large surface area.
- No mixer or aerators are required, thus conserving energy and operation cost, the gas produced can be collected and used.
- The UASB system is uncommonly simple and it does not require elaborate equipment and construction activities.

The most difficult problem with the UASB system is corrosion. Hence, all construction materials used to be carefully chosen. The UASB is not yet widely used in India. Its application for treatment of municipal wastewater is only 6 years old and the manipulates are not geared to adopt new technologies. However, its applications in industry are increasingly rapidly. Through the Ganga Action Plan, a beginning has been made which will be consolidated in the National River Action Plan.

9.4.2 Two-Stage, Aerobic Uni Tank System (TSU-System)

The two stage, aerobic Uni tank system and the tri-stage, anaerobic-aerobic Uni tank system with biological nitrogen removal (3SU - N System) has been developed in Europe. This is a cost-effective alternative to conventional activated sludge systems. The main advantages are reduction in capital and operational costs, flexible and reliable operation and high process performance. After the preliminary treatment (screening, gril removal equalization, no primary settling) the wastewater is first treated in a high loaded combined aeration-sedimentation stage. The BOD reduction is about 80-85%. The partially purified water then flows by gravity to a low loaded combined aeration-sedimentation stage where the residual BOD is removed to obtain a high quality effluent resulting in more than 98% removal of BOD. The advantages are list below.

- Less capital costs, No primary settling, Less total aeration volume, No separate sedimentation tanks, No sludge scraping, No sludge recycling facilities Rectangular tanks, compact construction possible, full use of available land, cheaper and easier to construct as compared to circular tanks, economical lengths of connecting pipes and channels, Compact system: smaller land area required.

- Less operational costs, less energy for aeration, No energy for sludge recycle, Less maintenance costs (less moving parts).
- Better process performance, High treatment efficiency, Control of sludge bulking, Simple and reliable process, reduced need for supervision.
- Easily controlled by microprocessor
- Flexible operation, Flexibility of temporary operation with half capacity, Restoration of full capacity without long time lag, Possible applications, brewing and malting wastewater treatment, Municipal wastewater treatment, Food processing wastewater treatment, Industrial wastewater treatment, Aerobic post treatment of anaerobic effluents from distilleries.

9.4.3 Root Zone Treatment

The process is a natural way of treating industrial or domestic wastes. The method developed in Sixties in Germany, is now commercialized for treatment of domestic and industrial wastewater, economically and efficiently. It has got three integrated components; reeds, reed bed and microbial organisms. In this system, contaminated water is allowed to flow underground through the root zones of especially designed reed beds. The reeds and the reed bed on the soil surface provide an efficient treatment system.

The reed bed serves as the host for more than 2000 species of bacteria and thousands of fungal species. These microbial organisms oxidize the organic matter both aerobically as well as an aerobically. Phosphate, sulphur and carbon compounds, nitrogenous materials reduce to their elemental forms. Heavy metals precipitate from solution and are bound into the soil matrix. Due to the high biodiversity of microbes, the root-zone system is capable of shock loads.

The root zone system is suitable for concentrations from a few mg/l upto 20,000 mg/l of COD & 4000 mg/l of nitrogen. It can be built for effluent throughout from about I m^3/day to more than 10,000 m^3/day. For domestic sewage the land requirement is around 0.2 m^2/person. But for the larger area requirement as compared to conventional methods, the root zone treatment system offers an ideal option for biological effluents because of its simplicity and ruggedness. Even in areas where land is a constraint, the system could be adopted with innovations like vertical treatment facility.

9.4.4 Submerged Aerobic Fixed Film (SAFF) Reactor

SAFF system is basically an aerobic process which incorporates two different biological growth systems viz., Attached Growth Process and Suspended Growth Process in a Single Reactor. Cross structured PVC Media is generally used as a medium for attached growth, where the media provides surface area in excess of 100m2/Cu. m volume. Large quantity of microbes starts growing inside the media uniformly. About 50 – 70% of the volume of the reactor is occupied by the PVC media for attached growth and the rest is utilized for suspended growth. Oxygen requirement of the system is met with the help of the fine bubble air diffusers provided at the bottom for which the necessary air is supplied through air blowers. The advantages are listed below:

- When compared with Conventional ASP system, SAFF system provides 5 – 10 times more surface area.
- It can be given as a package and skid mounted system.
- Sludge re-circulation is not required unlike conventional system where sludge recirculation is required to maintain the MLSS.
- SAFF system takes higher shock loads without reducing the plant performance because of large quantity of MLSS available inside the reactor.
- Rate of sludge generation is normally lesser as compared to conventional system.
- Higher loading of BOD on the media enables to reduce the aeration tank size.
- Fine bubble diffused aeration due to its increased oxygen transfer rates reduces the energy requirements as compared to surface aerators.
- Compactness of the plant encourages for indoor applications or basement applications.
- It is easy to operate & maintain and has no movable parts in this.

9.4.5 Fluidized Aerobic Bio – Reactor (FAB)

FAB reactor is an advanced wastewater treatment process utilizing free floating media which houses active biological cells. FAB is a hybrid process where both Attached Growth and Suspended Growth treatment Process simultaneously. Floating (Random) PVC Media is generally used as a medium for attached growth, where the media provides surface area in excess of 300m2/Cu. m volume. Large quantity of microbes starts growing on the media uniformly. About 35 – 50% of the volume of the reactor is occupied by the PVC media for attached growth and the rest is utilized for suspended growth. Oxygen requirement of the system is met with the help of the fine bubble air diffusers provided at the bottom for which the necessary air is supplied through air blowers. The advantages of FAB system are listed below:

- When compared with Conventional ASP system, this system provides more surface area hence less space requirement.
- It can be given as a package and skid mounted system.
- Sludge re-circulation is not required unlike conventional system where sludge recirculation is required to maintain the MLSS.
- FAB system takes higher shock loads without reducing the plant performance because of large quantity of MLSS available inside the reactor.
- Rate of sludge generation is normally lesser as compared to conventional system.
- Higher loading of BOD on the media enables to reduce the aeration tank size.
- Fine bubble diffused aeration due to its increased oxygen transfer rates reduces the energy requirements as compared to surface aerators.
- Compactness of the plant (very less foot print area) encourages for indoor applications or basement applications.
- It is easy to operate & maintain and has no movable parts in this.

9.5 ADVANCED WASTE WATER TREATMENT METHODS

The effluent from a typical secondary treatment plant still contains 20-40 mg/L BOD which may be objectionable in some streams. Suspended solids, in addition to contributing to BOD, may settle on the stream bed and inhibit certain forms of aquatic life. The BOD if discharged into a stream with low flow, can cause damage to aquatic life by reducing the dissolved oxygen content. In addition the secondary effluent contains significant amounts of plant nutrients and dissolved solids. If the waste water is of industrial origin, it may also contain traces of organic chemicals, heavy metals and other contaminants.

Different methods are used in advanced waste treatment to satisfy any of the several specific goals, which include the removal of 1. suspended solids 2. BOD 3. plant nutrients 4. dissolved solids and 5. toxic substances. These methods may be introduced at any stage of the total treatment process as in the case of industrial waterways or may be used for complete removal of pollutants after secondary treatment.

9.5.1 Removal of suspended solids

This treatment implies the removal of those materials that have been carried over from a secondary treatment settler. Many methods were proposed of which two methods were commonly used. The two methods are micro staining and chemical coagulation followed by settling and mixed media filtration.

Micro staining: It is a special type of filtration process which make use of filters oven from stainless steel wires with very fine pores of 60-70 microns size. These filters helps to remove very fine particles. High flow rates and low back pressures are achieved

Coagulation and flocculation: The object of coagulation is to alter these particles in such a way as to allow them to adhere to each other. Most colloids of interest in water treatment remain suspended in solution because they have a net negative surface charge that causes the particles to repel each other. The intended action of the coagulant is to neutralize that charge, allowing the particles to come together to form larger particles that can be more easily removed from the raw water.

The usual coagulant is alum $[Al_2(SO_4)_2 \bullet 18H_2O]$, though $FeCl_3$, $FeSO_4$ and other coagulants, such as polyelectrolytes, can be used. Alum when added to water, the aluminium in this salt hydrolyses by reactions that consume alkalinity in the water such as:

$$Al\ (HO)_6] + 3\ 3HCO_3 \text{ --- } Al(OH)_3(s) + 3CO_2 + 6H_2O \qquad .....(1)$$

The gelatinous hydroxide thus formed carries suspended material with it as it settles. Metal ions in coagulants also react with virus proteins and destroy upto 99% of the virus in water. Anhydrous ion (III) sulphate can also act as effective coagulant similar to aluminium sulfate. An advantage with iron (III) sulfate it that it works over a wide range of pH

Filtration: If properly formed, the addition of chemicals for promoting coagulation and flocculation can remove both suspended and colloidal solids. After the flocs are formed, the solution is led to a settling tank where the flocs are allowed to settle. While most of the flocculated material is removed in the settling tank, some floc do not settle. These flocs are removed by the filtration process, which is usually carried out using beds of porous media such as sand or coal. The current trend is to use a mixed - media filter which consists of fine garnet in the bottom layer, silica sand in the middle layer and coarse coal in the top layer which reduces clogging

Ultra Filtration

- Selectively filters only molecules of a specified size and weight.
- Removes e.g. various viruses.
- Used for sterilization, clarification, wastewater treatment.
- Membrane size 1 _ - 0.01 μm. is used

This is a dynamic filtering process with a predominance of physical (mechanical) phenomena in which chemical phenomena are also involved. The membranes used, polymeric or mineral, allow dissolved salts to pass while they reject high molecular weights selectively. The selectivity depends on the membrane structure and is defined as the cut-off of molecular weight, which the membrane can separate with an efficiency of 90 % (although this definition may not be rigorous depending on the molecular shape)

Commercial membranes applied in ultra filtering can separate substances with a molecular weight between 1.000 and 10.000. Ultra filtering systems generally work in a pressure range between 1.5 and 7 bar With industrial discharge waters the fluxes of permeate generally fluctuate between 0.5 and 1 - 5 m^3 / h / m^2 surface, depending on the concentration of the substances to be separated, with energy consumptions varying between 2 and 20 KWh per m3 of permeate. The single pass ultra filtering process is the simplest and most commonly used process for water treatment because it allows the recovery of high percentages of permeate (approximately 90-95 %).

There has been a relatively recent application of this technique in the metal finishing sector for the recovery of degreasing baths (the first cleaning bath in metal-finishing processes, for pieces which are still dirty with lubricating substances). The solution to be treated is passed through the membrane at a certain speed and under hydrostatic pressure, obtaining a concentrated fraction of oils and grease for disposal, while the filtrate is recovered and reused to prepare new baths.

Nano Filteration: The nano filtration technique is mainly used for the removal of two valued ions and the larger mono valued ions such as heavy metals. This technique can be seen as a coarse RO (reversed osmosis) membrane. Because nano filtration uses less fine membranes, the feed pressure of the NF system is generally lower compared to RO systems. Also the fouling rate is lower compared to Ro systems.

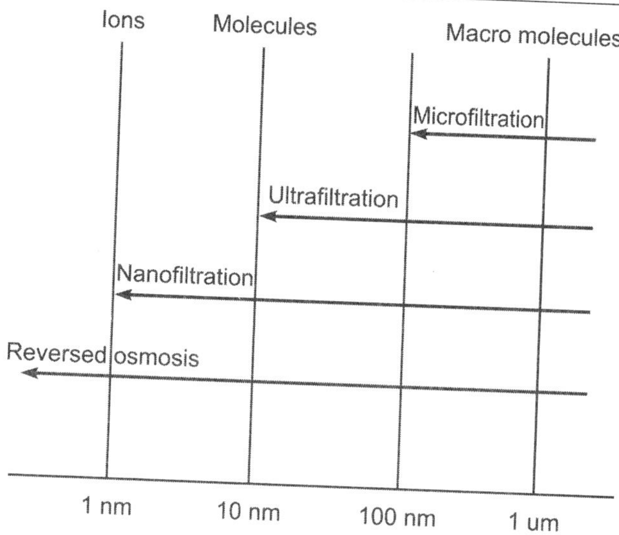

9.5.2 Removal of Dissolved Solids

The dissolved solids are of both organic and inorganic types. A number of methods have been investigated for the removal of inorganic constituents from waste water. Three methods which are finding wide application in advanced waste treatment are ion-exchange, electrodialysis and reverse osmosis. For the removal of soluble organics from waste water the most commonly used method is adsorption on activated carbon. Solvent extraction is also used to recover certain organic chemicals like phenol an d amines from industrial waste waters.

Ion exchange: This technique has been used extensively to remove hardness, and iron and manganese salts in drinking water supplies. It has also been used selectively to remove specific impurities and to recover valuable trace metals like chromium, nickel, copper, lead and cadmium from industrial waste discharges. The process takes advantage of the ability of certain natural and synthetic materials to exchange one of their ions.

A number of naturally occurring minerals have ion exchange properties. Among them the notable ones are aluminum silicate minerals, which are called zeolites. Synthetic zeolites have been prepared using solutions of sodium silicate and sodium aluminate. Alternatively synthetic ion-exchange resins composed of organic polymer with attached functional groups such as (strongly acidic cation exchange resins), or - COO - 3 -SO $H+^{-}$ H^{+} (weakly acidic cation exchange resins or $–N^{+}(CH_3)_3OH^{-}$ (strongly basic anion exchange resins) can be used.

In the water softening process, the hardness producing elements such as calcium and magnesium are replaced by sodium ions. A cation exchange resin in sodium form is normally used. The water-softening capability of cation exchange can be seen when sodium ion in the resin is exchanged for calcium ion in solution

Reverse osmosis: In the reverse osmosis process, de-mineralization water is produced by forcing water through semi permeable membranes at high pressure. In ordinary osmosis, if a vessel is divided by a semi permeable membrane (one that is permeable to water but not the dissolved material), and one compartment is filled with water and other with concentrated salt solution, water diffused through the membrane towards the compartment containing salt solution until the difference in water levels on the two sides of the membrane creates a sufficient pressure to counteract the original water flow. The difference in levels represents the osmotic pressure of the solution

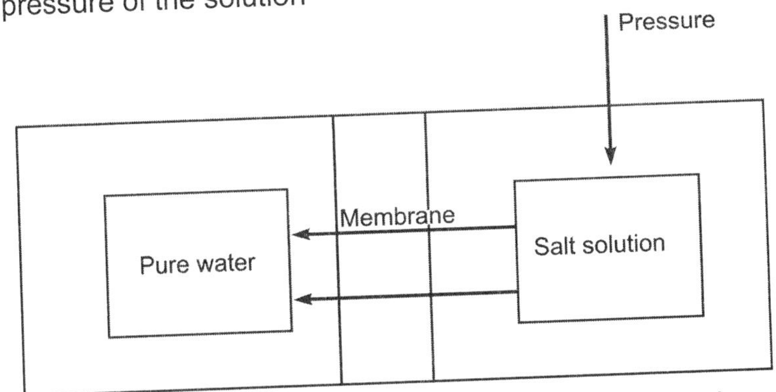

Industrial effluent treatment, using reverse osmosis, can be applied in the following main sectors

- Treatment of outflows containing colorings with their possible recovery.
- Treatment of outflows containing oily emulsions, latex and electro phoretic paints.
- Treatment of outflows from the metal-finishing industry with recovery of concentrated solutions of metal salts and reuse of the water in cleaning
- Treatment of waste water from organic chemical, in organic chemical and pharmaceutical industries

The application of reverse osmosis for wastewater treatment is significantly different from general process water purification. This is primarily due to the fact that wastewater generally contains higher levels and a more diverse range of contaminants. In addition, industrial wastewaters have a high degree of variability. Wastewaters vary from industry to industry and can change with hour to hour operation at any individual plant. The most important factor in treating industrial wastewater with RO is the against organic fouling, mineral scaling and chemical degradation. Before RO should even be considered, a complete cation/anion balance is required and possible floculants must be identified. Potential inorganic foulants and scalants of RO membranes include calcium, iron, aluminum, and other insoluble heavy metals. Possible organic foulants include surfactants, color bodies, flocculants, and bacteria. High BOD and COD levels can also contribute to membrane fouling.

A wide range of pretreatment technologies is available. Specifically in the metal finishing, printed circuit board and microelectronics industries, rinsewaters from fabrication operations are normally treated to remove heavy metals and are then discharged to the sewer. The effluent discharged to the sewer typically contains between 200 to 10,000 parts per million (ppm) total dissolved solids (TDS). With the proper pretreatment technology followed by RO, this effluent can be treated and recycled. Ion exchange treatment of the RO product water can further polish the water and make it suitable for all rinses. To design a successful and cost-effective system, it is necessary to evaluates each individual application because the pH, oxidizing potential and concentration of soluble salts of the wastewater effluents often exceed the operating limits of RO systems. After the detailed evaluation of the wastewater is complete, one need to determines the optimum preconditioning chemistry and selects the best pretreatment technology for the application.

Reverse Osmosis process generates high TDS waste stream reject. Approximately 25-40% of waste reject with high TDS concentration will be generated from feed water. This waste needs to be evaporated in forced evaporation systems to concentrate and remove the in organic impurities from it

9.5.3 Thermal Evaporation

Evaporation can take the form of vacuum distillation, atmospheric evaporation, and thermal evaporation. Vacuum distillation is accomplished by drawing a vacuum on a chamber and evaporating water at reduced temperatures, typically in the range of 90-150 degrees Fahrenheit. This technology is characterized by low energy cost, moderate to high manpower requirements, and very high capital cost. Atmospheric evaporation involves spraying the wastewater across a high surface area medium and blowing large volumes of air across the medium. This type of evaporation is characterized by moderate energy cost, moderate capital cost, high manpower requirements due to the tendency for fouling and reduced throughputs caused by changes in atmospheric conditions.

Thermal evaporation/distillation is accomplished by heating the wastewater to a boiling temperature and evaporating the waste stream at various rates based on the amount of energy (BTU's) input into the system. This type of evaporation is characterized by moderate to high energy cost, low manpower requirements, moderate capital cost, high flexibility and high reliability. This system has the ability to exhaust water as clean water vapor or recover water as distilled water. The advantages of Thermal Evaporation over Chemical Treatment are as follows:

Zero Discharge: Evaporation completely eliminates your discharge effluent. This eliminates accountability to your pollution control Board as well as the hassle and expense associated with potential discharge violations.

Total Solution: Chemical treatment does not completely address parameters such as emulsified oils, Chemical Oxygen Demand (COD), Biological Oxygen Demand (BOD), or dissolved solids in the discharge wastewater. This becomes more important each year as Pollution control discharge limits become increasingly strict

Lower Disposal Cost: Due to the addition of chemistry, the sludge volume being generated will be greater for chemical treatment compared to evaporation which typically does not require the addition of chemistry. This translates to lower disposal liability and cost for evaporation.

9.5.4 Removal of Dissolved Organic Compounds

One of the most commonly used techniques for removing organics involves the process of adsorption, which is the physical adhesion of chemicals on to the surface of the solid. The effectiveness of the adsorbent is directly related to the amount of surface area available to attract the particles of contaminant. The most commonly used adsorbent is a very porous matrix of granular activated carbon, which has an enormous surface area (~ 1000 m^2/g). Adsorption on activated carbon is perhaps the most economical and technically attractive method available for removing soluble organics such as phenols, chlorinated hydrocarbons, surfactants, and colour and odour producing substances from waste water.

Granular activated carbon treatment systems consist of a series of large vessels partially filled with adsorbent. Contaminated water enters the top of each vessel, trickles down through granulated activated carbon, and is released at the bottom. After a period of time, the carbon filter becomes clogged with adsorbed contaminants and must be either replaced or regenerated. Regeneration of the carbon is accomplished by heating it to 950 °C in a steam air atmosphere. This process oxidises surface, with an approximately 10% loss of carbon (Table 9.3).

Synthetic organic polymers such as Amberlite XAD-4 have hydrophobic surfaces and are quite useful in removing relatively insoluble organic compounds such as chlorinated pesticides. These absorbents are readily regenerated by solvents such as isopropanol and acetone

Table 9.3 Water-quality based decisions

Agriculture	**Forestry**
Animal waste management	Ground cover maintenance
Conservation tillage	Limiting disturbed areas
Contour farming	Log removal techniques
Strip-coping	Pesticide and herbicide management
Cover crops	Proper management of roads
Crop rotation	Removal of debris
Fertilizer management	Riparian zone management
Integrated pest management	**Mining**
Rang and pasture management	Underdrains
Terraces	Walter division
Construction	**Multicategory**
Limiting disturbed areas	Buffer vegetated strips
Non-vegetative soil stabilization	Detention and sedimentation basins
Runoff detention and retention	Devices to encourage infiltration
Urban	Vegetated waterways
Blood storage	Interception and diversion of runoff
Porous pavements	Sediment traps
Runoff detention and retention	Streamside management zones
Street cleaning	Vegetative stabilization and mulching

Source: Guidance for water Quality-Based Decisions: The TMDL Process USEPA, 1991/EPA440-4-91-001

9.6 AIR POLLUTION CONTROL TECHNOLOGIES

Air pollution is the introduction of chemicals, particulate matter, or biological materials that cause harm or discomfort to humans or other living organisms, or damages the natural environment into the atmosphere. The atmosphere is a complex dynamic natural gaseous system that is essential to support life on planet Earth. Stratospheric ozone depletion due to air pollution has long been recognized as a threat to human health as well as to the Earth's ecosystems. Schematic drawing, causes and effects of air pollution: 1. greenhouse effect, 2. particulate contamination, 3. increased UV radiation, 4. acid rain, 5. increased ozone concentration, 6. increased levels of nitrogen oxides An air pollutant is known as a substance in the air that can cause harm to humans and the environment. Pollutants can be in the form of solid particles, liquid droplets, or gases. In addition, they may be natural or man-made.[1] Pollutants can be classified as either primary or secondary. Usually, primary pollutants are substances directly emitted from a process, such as ash from a volcanic eruption, the carbon monoxide gas from a motor vehicle exhaust or sulfur dioxide released from factories.

Secondary pollutants are not emitted directly. Rather, they form in the air when primary pollutants react or interact. An important example of a secondary pollutant is ground level ozone — one of the many secondary pollutants that make up photochemical smog. Note that some pollutants may be both primary and secondary: that is, they are both emitted directly and formed from other primary pollutants. About 4 percent of deaths in the United States can be attributed to air pollution, according to the Environmental Science Engineering Program at the Harvard School of Public Health.

9.6.1 Control Devices

The following items are commonly used as pollution control devices by industry or transportation devices. They can either destroy contaminants or remove them from an exhaust stream before it is emitted into the atmosphere.

Particulate control

- Mechanical collectors (dust cyclones, multicyclones)
- Electrostatic precipitators an electrostatic precipitator (ESP), or electrostatic air cleaner is a particulate collection device that removes particles from a flowing gas (such as air) using the force of an induced electrostatic charge. Electrostatic precipitators are highly efficient filtration devices that minimally impede the flow of gases through the device, and can easily remove fine particulate matter such as dust and smoke from the air stream.
- Bag houses Designed to handle heavy dust loads, a dust collector consists of a blower, dust filter, a filter-cleaning system, and a dust receptacle or dust removal system (distinguished from air cleaners which utilize disposable filters to remove the dust).

- Particulate scrubbers Wet scrubber is a form of pollution control technology. The term describes a variety of devices that use pollutants from a furnace flue gas or from other gas streams. In a wet scrubber, the polluted gas stream is brought into contact with the scrubbing liquid, by spraying it with the liquid, by forcing it through a pool of liquid, or by some other contact method, so as to remove the pollutants.

Scrubbers

- Baffle spray scrubber
- Cyclonic spray scrubber
- Ejector venturi scrubber
- Mechanically aided scrubber
- Spray tower
- Wet scrubber

NOx control

- Low NOx burners
- Selective catalytic reduction (SCR)
- Selective non-catalytic reduction (SNCR)
- NOx scrubbers
- Exhaust gas recirculation
- Catalytic converter (also for VOC control)

VOC abatement

- Adsorption systems, such as activated carbon
- Flares
- Thermal oxidizers
- Catalytic oxidizers
- Biofilters
- Absorption (scrubbing)
- Cryogenic condensers
- Vapor recovery systems

Acid Gas/SO_2 control

- Wet scrubbers
- Dry scrubbers
- Flue gas desulfurization

Mercury control

- Sorbent Injection Technology
- Electro-Catalytic Oxidation (ECO)
- K-Fuel

Miscellaneous associated equipment

- Source capturing systems
- Continuous emissions monitoring systems (CEMS)

9.7 SOIL POLLUTION CONTROL TECHNOLOGY

Soil pollution is defined as the build-up in soils of persistent toxic compounds, chemicals, salts, radioactive materials, or disease causing agents, which have adverse effects on plant growth and animal health. Soil is the thin layer of organic and inorganic materials that covers the Earth's rocky surface. The organic portion, which is derived from the decayed remains of plants and animals, is concentrated in the dark uppermost topsoil. The inorganic portion made up of rock fragments, was formed over thousands of years by physical and chemical weathering of bedrock. Productive soils are necessary for agriculture to supply the world with sufficient food. There are many different ways that soil can become polluted, such as:

- Seepage from a landfill
- Discharge of industrial waste into the soil
- Percolation of contaminated water into the soil
- Rupture of underground storage tanks
- Excess application of pesticides, herbicides or fertilizer
- Solid waste seepage

The most common chemicals involved in causing soil pollution are:

- Petroleum hydrocarbons
- Heavy metals
- Pesticides
- Solvents

Types of Soil Pollution

- Agricultural Soil Pollution
  (i) pollution of surface soil
  (ii) pollution of underground soil

- Soil pollution by industrial effluents and solid wastes
  - (i) pollution of surface soil
  - (ii) disturbances in soil profile
- Pollution due to urban activities
  - (i) pollution of surface soil
  - (ii) pollution of underground soil

9.7.1 Control of Soil Pollution

The following steps have been suggested to control soil pollution. To help prevent soil erosion. In general we would need less fertilizer and fewer pesticides if we could all adopt the three R's: Reduce, Reuse, and Recycle. This would give us less solid waste.

Reducing chemical fertilizer and pesticide use: Applying bio-fertilizers and manures can reduce chemical fertilizer and pesticide use. Biological methods of pest control can also reduce the use of pesticides and thereby minimize soil pollution.

Reusing of materials: Materials such as glass containers, plastic bags, paper, cloth etc. can be reused at domestic levels rather than being disposed, reducing solid waste pollution.

Recycling and recovery of materials: This is a reasonable solution for reducing soil pollution. Materials such as paper, some kinds of plastics and glass can and are being recycled. This decreases the volume of refuse and helps in the conservation of natural resources. For example, recovery of one tonne of paper can save 17 trees.

Reforesting: Control of land loss and soil erosion can be attempted through restoring forest and grass cover to check wastelands, soil erosion and floods. Crop rotation or mixed cropping can improve the fertility of the land.

1. Use of pesticides should be minimized.
2. Use of fertilizers should be judicious.
3. Cropping techniques should be improved to prevent growth of weeds.
4. Special pits should be selected for dumping wastes.

5. Controlled grazing and forest management.
6. Wind breaks and wind shield in areas exposed to wind erosin
7. Planning of soil binding grasses along banks and slopes prone to rapid erosin.
8. Afforestation and reforestation.

9.7.2 Solid Waste Treatment

Proper methods should be adopted for management of solid waste disposal. Industrial wastes can be treated physically, chemically and biologically until they are less hazardous. Acidic and alkaline wastes should be first neutralized; the insoluble material if biodegradable should be allowed to degrade under controlled conditions before being disposed. As a last resort, new areas for storage of hazardous waste should be investigated such as deep well injection and more secure landfills. Burying the waste in locations situated away from residential areas is the simplest and most widely used technique of solid waste management. Environmental and aesthetic considerations must be taken into consideration before selecting the dumping sites. Incineration of other wastes is expensive and leaves a huge residue and adds to air pollution. Pyrolysis is a process of combustion in absence of oxygen or the material burnt under controlled atmosphere of oxygen. It is an alternative to incineration. The gas and liquid thus obtained can be used as fuels. Pyrolysis of carbonaceous wastes like firewood, coconut, palm waste, corn combs, cashew shell, rice husk paddy straw and saw dust, yields charcoal along with products like tar, methyl alcohol, acetic acid, acetone and a fuel gas.

Anaerobic/aerobic decomposition of biodegradable municipal and domestic waste is also being done and gives organic manure. Cow dung which releases methane into the atmosphere, should be processed further in 'gobar gas plants' to produce 'gobar gas' and good manure.

9.8 PHYTOREMEDIATION TECHNOLOGIES

Phytoremediation refers to the use of plants to remove pollutants from soil and groundwater, or to assist in the degradation of contaminants to a less toxic form. Some plants are able to extract and concentrate particular elements from the environment, thereby offering a permanent means of remediation. The plant tissue, which is rich in accumulated contaminants, can be harvested and safely processed. Remediation also occurs when bacteria on the roots of the plant degrade pollutant species, or when the roots draw contaminated ground moisture closer to the surface, exposing contaminant species to microbes in a higher oxygen-containing environment. Some of the techniques are presented in this section. They are: phytoextraction, phytostabilization, phytostimulation, phytotransformation, rhizofiltration, constructed wetlands.

9.8.1 Phytoextraction

The use of plants to remove contaminants from the environment and concentrate them in above-ground plant tissue is known as phytoextraction.

Applicability: Phytoextraction was primarily employed to recover heavy metals from soils, however, this technology is now applicable to other materials in different media. Greenhouse-based hydroponic systems using plants with high contaminant root uptake and poor translocation to the shoots are currently being researched for removal of heavy metals and radionuclides from water .These plants also are referred to as hyperaccumulators. Plants with high growth rates (>3 tons dry matter/hectare-year) and the ability to tolerate high metal concentrations in harvestable parts of the plants (>1,000 mg/kg) are needed for practicable treatment .

Limitations: Effective extraction of toxic metals by hyperaccumulators is limited to shallow soil depths of up to 24 inches. If contamination is at substantially greater depths (e.g., 6 to 10 feet), deep-rooted poplar trees can be used, however, there is concern about leaf litter and associated toxic residues.

Despite having amiable metal-accumulating characteristics, currently available hyperaccumulators lack suitable biomass production, physiological adaptability to varying climatic conditions, and adaptability to current agronomic techniques.

9.8.2 Phytostabilization

Phytostabilization involves the reduction of the mobility of heavy metals in soil. Immobilization of metals can be accomplished by decreasing wind-blown dust, minimizing soil erosion, and reducing contaminant solubility or bioavailability to the food chain. The addition of soil amendments, such as organic matter, phosphates, alkalizing agents, and biosolids can decrease solubility of metals in soil and minimize leaching to groundwater. The mobility of contaminants is reduced by the accumulation of contaminants by plant roots, absorption onto roots, or precipitation within the root zone. In some instances, hydraulic control to prevent leachate migration can be achieved because of the large quantity of water transpired by plants.

Applicability: The use of phytostabilization to keep metals in their current location is particularly attractive when other methods to remediate large-scale areas having low contamination are not feasible. Remediation is difficult at locales having high metals concentration because of soil toxicity. Plants should be able to tolerate high levels of contaminants, have high production of root biomass with the ability to immobilize contaminants, and the ability to hold contaminants in the roots.

Limitations: Phytostabilization is useful at sites with shallow contamination and where contamination is relatively low. Plants that accumulate heavy metals in the roots and in the root zone typically are effective at depths of up to 24 inches. Metals that are readily translocated to leaves in plants may limit the applicability of phytostabilization due to potential affects to the food chain.

9.8.3 Phytostimulation

Phytostimulation, also referred to as enhanced rhizosphere biodegradation, rhizodegradation, or plant-assistedbioremediation /degradation, is the breakdown of organic contaminants in the soil via enhanced microbial activity in the plant root zone or rhizosphere. Microbial activity is stimulated in the rhizosphere in several ways: 1. compounds, such as sugars, carbohydrates, amino acids, acetates, and enzymes, exuded by the roots enrich indigenous microbe populations; 2. root systems bring oxygen to the rhizosphere, which ensures aerobic transformations; 3 fine-root biomass increases available organic carbon; 4. mycorrhizae fungi, which grow within the rhizosphere, can degrade organic contaminants that cannot be transformed solely by bacteria because of unique enzymatic pathways; and 5. the habitat for increased microbial populations and activity is enhanced by plants .

Applicability: This method is useful in removing organic contaminants, such as pesticides, aromatics, and polynuclear aromatic hydrocarbons (PAHs), from soil and sediments .Chlorinated solvents also have been targeted at demonstration sites.

Limitations: Locations at which phytostimulation is to be implemented should have low levels of contamination in shallow areas. High levels of contaminants can be toxic to plants.

9.8.4 Phytotransformation

Phytotransformation, also referred to as phytodegradation, is the breakdown of organic contaminants sequestered by plants via: (1) metabolic processes within the plant; or (2) the effect of compounds, such as enzymes, produced by the plant. The organic contaminants are degraded into simpler compounds that are integrated with plant tissue, which in turn, foster plant growth .Remediation of a site by phytotransformation is dependent on direct uptake of contaminants from the media and accumulation in the vegetation. The release of volatile contaminants to the atmosphere via plant transpiration, called phytovolatilization, is a form of phytotransformation. Although transfer of contaminants to the atmosphere may not achieve the goal of complete remediation, phytovolatilization may be desirable in that prolonged soil exposure and the risk of groundwater contamination are reduced.

Applicability: Phytotransformation can be employed to remediate sites contaminated with organic compounds. Certain enzymes produced by plants are able to breakdown and convert chlorinated solvents (e.g., trichloroethylene), ammunition wastes, and herbicides. This technology also can be used to remove contaminants from petrochemical sites and storage areas, fuel spills, landfill leachates, and agricultural chemicals. Successful implementation of this technology requires that the transformed compounds that accumulate within the plant be non-toxic or significantly less toxic than the parent compounds. In some applications, phytotransformation may be used in concert with other remediation technologies or as a polishing treatment.

Limitations: This technology usually requires more than one growing season to be efficient. Soil must be less than 3 ft in depth and groundwater within 10 ft of the surface. Contaminants may still enter the food chain through animals or insects that eat plant material. Soil amendments may be required, including chelating agents to facilitate plant uptake by breaking bonds binding contaminants to soil particles.

9.8.5 Rhizofiltration

Rhizofiltration refers to the use of plant roots to absorb, concentrate, and precipitate toxic metals from contaminated groundwater. Initially, suitable plants with stable root systems are supplied with contaminated water to acclimate the plants. These plants are then transferred to the contaminated site to collect the contaminants, and oncc the roots are saturated, they are harvested. Rhizofiltration allows in-situ treatment, minimizing disturbance to the environment.

Applicability: A suitable plant for rhizofiltration applications can remove toxic metals from solution over an extended period of time with its rapid-growth root system. Various plant species have been found to effectively remove toxic metals such as Cu (2+), Cd (2+), Cr (6+), Ni (2+), Pb (2+), and Zn (2+) from aqueous solutions .Low level radioactive contaminants also can be removed from liquid streams .

Limitations: Rhizofiltration is particularly effective in applications where low concentrations and large volumes of water are involved. Plants that are efficient at translocating metals to the shoots should not be used for rhizofiltration because more contaminated plant residue is produced.

9.8.6 Constructed Wetlands

Constructed wetlands are engineered, man-made ecosystems specifically designed to treat wastewater, mine drainage, and other waters by optimizing the biological, physical, and chemical processes that occur in natural wetland systems. Constructed wetlands can provide effective, economical, and environmentally-sound treatment of wastewater as well as serve as wildlife habitats.

Constructed wetland systems are grouped into three main types: free-water surface (FWS), subsurface flow systems (SFS), or aquatic plant systems (APS). FWS systems, or soil substrate systems, consist of aquatic plants rooted in a soil substrate within a constructed earthen basin that may or may not be lined depending on soil permeability and groundwater protection requirements . FWS systems are designed to accept preliminary-treated, low-velocity wastewater, in plug flow, over the top of the soil media or at a depth between 1 and 18 inches. SFS are typically gravel substrate systems that are similar to FWS systems, however, aquatic vegetation is planted in gravel or crushed stone and wastewater flows approximately 6 inches below the surface of the media. The aggregate typically has a depth between 12 and 24 inches. No visible surface flow is evident in SFS .APS also are similar to FWS systems, but the water is located in deeper ponds and aquatic floating aquatic plants or submerged plants are used.

Applicability: Constructed wetlands may be used to treat municipal wastewater, agricultural runoff, mine drainage, and other effluents. Biochemical oxygen demand (BOD) and total suspended solids (TSS) are effectively reduced by these man-made wetland systems.

Limitations: Technical guidance for designing and operating constructed wetlands may be limited due to the lack of long-term operational data. Potential seasonal variability and impact on wildlife may negatively impact system operation and securing of permits, respectively . Relatively large parcels of land are required and water consumption is high due to large evapotranspiration rates.

9.9 BIOREMEDIATION TECHNOLOGIES

Bioremediation can be defined as any process that uses microorganisms, fungi, green plants or their enzymes to return the natural environment altered by contaminants to its original condition. Bioremediation may be employed to attack specific soil contaminants, such as degradation of chlorinated hydrocarbons by bacteria. An example of a more general approach is the cleanup of oil spills by the addition of nitrate and/or sulfate fertilisers to facilitate the decomposition of crude oil by indigenous or exogenous bacteria.

Bioremediation can occur on its own (natural attenuation or intrinsic bioremediation) or can be spurred on via the addition of fertilizers to increase the bioavailability within the medium (biostimulation). Bioremediation technologies can be generally classified as *in situ* or *ex situ*. *Insitu* bioremediation involves treating the contaminated material at the site while *exsitu* involves the removal of the contaminated material to be treated elsewhere. Some examples of bioremediation technologies are: bioaugmentation, biostimulation, bioreactors, land-based treatments, fungal Remediation.

9.9.1 Bioaugmentation

Bioaugmentation refers to the introduction of specially selected or genetically engineered strains of microbes to a contaminated site. If site assessments reveal that species of indigenous microorganisms are unable to degrade target contaminants, exogenous microorganisms with the required biochemical capabilities can be introduced to successfully degrade specific wastecompounds.

9.9.2 Biostimulation

Biostimulation refers to the addition of oxygen and/or inorganic nutrients to indigenous microbial populations in soils and groundwater. *In situ* or *ex situ* methods can be employed to stimulate biodegradation of contaminants.

9.9.3 Bioreactors

Bioreactors represent highly controlled methods of treating contaminated soils and groundwater. Because temperature, pH, nutrient levels, and agitation can be controlled in constructed batch- or continuously-fed reactors, microbial activity, and thus contaminant degradation, can be optimized.

9.9.4 Land-based Treatments

Land-based treatment, or solid-phase remediation, of contaminated soil usually involves *ex situ* remediation methods. Excavated soil can be treated in piles or in constructed treatment cells.

9.9.5 Fungal Remediation

White-rot fungus, *phaneorochaete chrysosporium*, can bind to, and in some instances, mineralize a wide array of organic pollutants, including polychlorinated biphenyls (PCBs), polycyclic aromatic hydrocarbons (PAHs), and the predominant conventional explosives TNT, RDX, and HMX. The lignin-degrading, or wood-rotting, enzymes produced by white-rot fungus has been reported to be the key to its degradation capabilities. Two different treatment configurations have been tested for white-rot fungus -- bioreactor and *in situ* systems.

9.10 HAZARD WASTE TREATMENT (HWT)

Due to their high potential for public health and environmental damage, some hazardous wastes require special control procedures. Management of these wastes means awareness and control over them from the time of generation through their transportation, temporary storage, treatment and disposal. Effective identification and labeling of wastes by the generators is essential to the effective operation of any manifest-based system. In addition management of hazardous waste means more than careful disposal. It implies consideration of alternate methods and schemes, both institutional and technical, to hazardous waste reduction.

The first priority is to find ways to eliminate uses of hazardous substances altogether. Elimination might be achieved by changing manufacturing processes or by substituting products that can satisfy the same need without creating hazardous wastes (an example would be the substitution of concrete posts in place of, creosote-treated wooden posts).

The next priority is to reduce the amount generated. Again manufacturing processes can be modified to yield substantial waste reduction. For example changes in the chemical reaction conditions can minimize production of by-product hazardous substances. In some instances potentially hazardous catalysts, can be replaced by catalysts that are non hazardous or that can be recycled rather than discarded.

The third strategy is to recycle hazardous substances such as solvents and acids to maximise their use before treatment and disposal becomes necessary. Many industrial processes that use solvents are equipped for solvent recycle for reasons of both economics and pollution control. A number of operations are used in solvent purification. Entrained solids are removed by settling, filtration, or centrifugation. Drying agents may be used to remove water. Fractional distillation is used to separate solvent from impurities, water and other solvents.

Finally, hazardous substances can be **treated** to reduce their volume and toxicity. Eliminating, reducing, and recycling hazardous substances reduces treatment and disposal costs and often the generators can find it in their own economic best interest to actively seek out these source reduction opportunities as a way to reduce the overall cost of doing business.

9.10.1 HWT Technologies

Even with after vigorous hazardous waste reduction program, there will still be large quantities of hazardous wastes that will require treatment and disposal. The treatment technologies have been categorised as physical, chemical, biological, thermal, or stabilisation/fixation.

Physical treatment processes include gravity separation, phase change systems, such as air and steam stripping of volatiles from liquid wastes, and various filtering operations, including carbon adsorption.

Chemical treatment transforms waste into less hazardous substances using such techniques as pH neutralisation, oxidation or reduction, and precipitation. Biological treatment uses microorganisms to degrade organic compounds in the waste stream.

Thermal destruction processes include incineration, which is increasingly becoming a preferred option for the treatment of hazardous wastes, and pyrolysis, which is the chemical decomposition of waste is brought about by heating the material in the absence of oxygen.

Stabilisation techniques involve removal of excess of water from a waste and solidifying the remainder either by mixing it with a stabilising agent such as portland cement, or vitrifying it to a glassy substance.

Most treatment measures have both physical and chemical aspects. The appropriate treatment technology for the hazardous wastes depends on thc nature of the wastes. The type of physical treatment to be applied to wastes depends strongly upon the physical properties of the material treated, including the state of matter, solubility in water and organic solvents, density, volatility, boiling point and melting point.

9.10.2 Physical Treatment Methods

Various physical treatment methods adopted in industry are: adsorption, resin adsorption, sedimentation, electrodialysis, reverse osmosis, solvent extraction, distillation, evaporation and filtration.

Adsorption: Adsorption on activated carbon occurs when a molecule is brought up to its surface and held there by physical and /or chemical forces. This process is reversible, thus allowing activated carbon to be regenerated and reused by proper application of heat and steam, or solvent. The factors that relate to adsorption capacity are:

- Greater surface area produces greater adsorption capacity [e.g: Activated carbon has large surface area (500-1500 m^2/g)]
- Adsorptivity increases as the solubility of the solute (in solvent) decreases. Thus, for hydrocarbons, adsorption increases with molecular weight
- For solutes with ionisable groups, maximum adsorption will be achieved at a pH corresponding to minimum ionisation.
- Adsorption capacity decreases with increasing temperature.

One additional point to be noted is that biological activity usually takes place in a carbon bed. If the concentration of the adsorbed species is high enough and the material is biodegradable and non toxic to the bacteria, then biological activity may significantly increase the effective removal capacity. Removal through adsorption by activated carbon has been applied to non aqueous waste stream such as petroleum fraction, syrups, vegetable oils, and pharmaceutical preparations. Colour removal is the most common objective in such cases. Current waste treatment applications are limited to aqueous solutions.

Resin adsorption: Waste treatment by resin involves two basic steps: 1. contacting the liquid waste stream with resin and allowing the resin to adsorb the solutes from the solution; and 2. subsequently regenerating the resins by removing the adsorbed chemicals, by simply washing with proper solvent.

The adsorption of a nonpolar molecule on to a hydrophobic resin (e.g., styrene divinyl-benzene-based resin) results primarily from the effect of Vander Waal's forces. In other cases other type of interactions such as dipole-dipole interaction and hydrogen bonding are important. In a few cases ion-exchange mechanism may be involved. For the removal of organic dye wastes from water, two different resins were employed: In this case the waste stream is first contacted with a normal polymeric adsorbent and then with an ion exchange resin.

Sedimentation : Sedimentation is a physical process where by particles suspended in a liquid settle by means of gravity. The fundamental elements of most sedimentation processes are:

- a basin or container of sufficient size to maintain the liquid to be treated in a relatively quiescent state for a specified period of time
- a means of directing the liquid to be treated into the above basin in a manner conducive to settling.
- a means of physically removing the settled particles from the liquid.

Sedimentation can be either a batch or a continuous process. Continuous processes are by far the most common, particularly when large volumes of liquid are to be treated. This technique has been widely used in the removal of heavy metals from iron and steel industry waste water; removal of fluoride from aluminium production waste water; and removal of heavy metals from waste water from copper smelting and from metal finishing industry and waste water stream from organic chemicals

Electrodialysis: The electrodialysis involves the separation of an aqueous stream (more concentrated in electrolyte than the original) and a depleted stream. Success of the process depends on special synthetic membranes, usually based on ion-exchange resins, which are permeable only to a single type of ion. Cation exchange membranes permit passage only of positive ions, under the influence of electric field, while anion exchange membranes permit passage only of negatively charged ions.

The feed water is passed through compartments formed by the spaces between alternating cation-permeable and anion permeable membranes held in a stack. At each end of the stack is an electrode having the same area as the membranes. A dc potential applied across the stack causes the positive and negative ions to migrate in opposite directions.

This technique has been used for desalination to produce potable water from brackish well water. In food industry electrodialysis is used for desalting whey. The chemical industry uses this technique for enriching or depleting solutions, and for removing mineral constituents from product streams.

Reverse osmosis: This technique which is most widely used consists of a membrane permeable to solvent but impermeable to most dissolved species, both organic and inorganic. These devices use pressure to force the contaminated water against the semipermeable membrane. The membrane acts as a filter, allowing the water to be pushed through the pores, but restricting the passage of larger molecules that are to be removed.

Cellulose acetate membranes were used in the past, but nowadays polysulphones and polyamides are increasingly popular for use at high pH values. Because of the susceptibility of the membranes to chemical attack and fouling, and the susceptibility of the flow system to plugging and erosion, it is common to preprocess the feed water to remove oxidising materials. The reverse osmosis technique has been widely used for desalination of sea or brackish water.

It has also been successfully used in the treatment of electroplating rinse waters, not only to meet effluent discharge standards, but also to recover concentrated metal salt solutions for reuse. It has also been used for treatment of waste stream from paper and food processing industries.

Solvent extraction: Solvent extraction is the separation of the constituents of a liquid solution by contact with another immiscible liquid. If the substances comprising the original solution distribute themselves differently between the two liquid phases, a certain degree of separation will result and this may be enhanced by the use of multiple contacts. The major application of solvent extraction to waste treatment has been in the removal of phenol from by-product water produced in coal coking, petroleum refining, and chemical synthesis that involve phenol.

The use of supercritical fluids (SCFs) most commonly CO_2 as extraction solvent, has been one of the more promising approaches to solvent extraction. SCFs are fluids existing at or above the lowest temperature at which condensation may occur. Above the critical temperature certain fluids exhibit characteristics that enhance their solvent properties. Organic materials, which are only slightly soluble in particular solvents at room temperature, become completely miscible with the solvent when under supercritical conditions. The excellent solvent properties result from the rapid mass transfer ability and the very low density that characterises an SCF. Major advantages of SCFs are short residence times with no char formation. Some of the important application of these SCFs, has been in the extraction of organohalide pesticide from soil, extraction of oil from emulsions used in aluminium and steel processing, and regeneration of spent activated carbon. Supercritical ethane has been used to purify waste oils contaminated with PCBs, metals and water.

Distillation: Distillation is expensive and energy intensive and can probably be justified only in cases where valuable product recovery is feasible (e.g., solvent recovery). This technique has only limited application in the treatment of dilute aqueous hazardous wastes.

Evaporation: Evaporation process is used for the treatment of hazardous waste such as radioactive liquids and sludges and concentrating of plating

available about 4019.5 sq.m for built up area like production blocks, raw material stores, finished goods godowns, utilities, R&D, QC, administrative block and pollution control facilities. About 1550 Sq.m for Roads, 2395.5 Sq.m for Vacant area and 4300 sq.m greenbelt area.

The proposed green belt at the .project site will form an effective barrier between the plant and the surroundings. Open spaces, where tree plantation may not be possible, will be covered with shrubs and grass to prevent erosion of topsoil. Adequate attention has been paid to plantation of trees, their maintenance and protection based on the geology, soil condition and topography of the site area.

Green belt will be developed around the plant site, whatever space is available around the periphery of the plant will be planned to be utilized for green belt. Other open spaces within the factory will be converted to green areas in the form of lawns or flowering plants. A wide range of plant species wilt be planted in and around the premises to help capture the fugitive emissions and noise levels from the plant premises. This wide range covers plants of fast growing type with thick canopy cover, perennial green nature, native origin and a large leaf area index. A specialist in horticulture may be appointed to identify any other native species and also supervise greenbelt development.

9.12.1 Advantage of Green Belt Development

- The biological activity of the particles at various locations necessarily vary because of difference of pollutant source profiles. These variations are expressions of both quantitative and qualitative differences, as for instance the relative amount of sulfuric acid mist, sulfates, or other reactive substances in the particulate mix or the relative amounts of specific carcinogenic compounds in the organic fraction of airborne particulate.
- It, therefore, can be seen that the evaluation of biological activity ascribable to "particulate" is complex and depends not only on the total quantity, size range and intrinsic physical or chemical properties, but also on their chance for interaction in the polluted air. The opportunity for variation in biological activity is enormous.
- Stomata are microscopic pores on the underside (abaxial) of the leaf. These stomata allow the plant leaves takes in Carbon Dioxidë (CO_2) and lets out Oxygen (O_2), and also allows water vapor out in the process of transpiration. As air passes through the stomata, most of the airborne particles will not pass through the stomata but will rather land on the' leaf's outer surface. This is similar to a filter, where air is pulled through the filter by an air pump and the airborne particles

deposit on the filter surface. If this air flow is the major cause of particles depositing on the leaf, the result will be that the concentration of particles on the abaxial surface of the leaf will be higher than that of the top surface (adaxial) because the airflow through the stomata will be pulling more particles onto the bottom surface.

- There is a certain amount of force needed for particles to stick to a surface. This amount is greater depending on the size of the particles. Because the airflow through the stomata is not very powerful, only the smaller particles will stick to the bottom surface. The particles on the top surface of the leaves will mainly be from the settling of dust. Because settled particles are mostly larger ones, those found on the top surface will be mostly larger. Therefore, analysis of the particle sizes on the leaves will show that particles on the tops of leaves are on average, larger than those on the bottom of leaves.
- Different types of leaves tend to have differences in several aspects of their surfaces. Some types of leaves have greater surface rigidity or roughness than other leaves, which may affect their stickiness or particle solubility. Stickier leaves would be better for collecting particles because more particles would stick fo their surface. Therefore, some types of leaves may be better for use in this type of analysis than others.
- It has been derived that trees can delimit the fine particulate pollution and have tremendous potential for improved air quality with substantial cost savings. This study will help to quantify the relative ability of individual tree species for removing fine particulates such as PM25. The plantation of urban trees can be evaluated in terms of money saved vis-à-vis expenditure involved in implantation of fine particulate strategies.
- Trees can act as efficient biological filters, removing significant amounts of particulate pollution from urban atmospheres. The study indicated that three has been significant difference in interception of particulate matter ($PM_{2.5}$) by different tree species.
- It is recommended based on the studies that Green cover /areas of Highly Dust capturing plant species should be developed around residential areas / industrial area, since dust capturing plants species can act as efficient biological filters, removing significant amounts of particulate pollution from urban atmospheres. The dust capturing phenomenon of plant species is a cost effective technology for reduction of particulate load in urban agglomerations.

The geographical, environmental, morphological, anatomical & physiological aspects of plants species have been found influencing the dust capture by plant species, therefore following criteria should be adopted for selection of plant species for green belt development in urban areas.

- The species should be adapt to site and should be able to produce optimum harvest on a sustained basis for example tree like Ficus religiosa (Peepal), Ficus bengalensis (Banyan), Ficus elastica (Indian Rubber) and Artocarpus integrifolia (Jack Fruit).
- The leaf litter should decompose quickly thus adding organic matter to the soil tree like Acacia farmesiana (Vilayati kikar), Delonix regiosa (Gulmohar), Accacia nelotica (Babul), Azadirachta indica (Neem) Melia azadirachta (Melia) are suitable for the purpose.
- The species should preferably be capable of enriching soil, through nitrogen fixation or any other mechanism tree like members of Leguminaceare family such as Luceana leucophloea (Shoe babool), Acacia farmesiana (Vilayati kikar) are better nitrogen fixing capabilities.
- The morphological characters of the species must suit the objectives of plantation and the cultivation practice; e.g. a wide crown may be preferred for dust capturing and fuel wood plantation but small-narrow crown with minimum effect on agriculture crop and providing valuable wood.
- Multi-purpose tree plant species have a special significance in fulfilling the objectives of environment as well as needs of the people. The combination of species to address the local needs are more beneficial. The trees like Quaking Aspen (Populus tremuloides); Blue Gum (Eucalyptus globules: Acacia farmesiana (Vilayati kicker), Delonix regiosa (Gulmohar), Accacia nelotica (Babul), Azadirachta indica (Neem) melia Azadirachta (Melia) are more valuable.
- The tree products should have acceptable characteristics to suit local customs and traditions flowering Herbs & shrubs species like Grape Jasmine (Gardenia jasminoides) Crown Daisy (Chrysanthamum species) Lily (Lillium species) ; Sunflower (Helianthus annuus) etc;

9.12.2 Design of Green Belt

Project site area is covered with sheet rock. Developing the greenbelt in this project needs scientific approach. As far as possible the following guidelines will be considered in green belt development programme.

All around the site, a green belt of 3 m width will be developed as bio-fence and bio-defense. In view of the presence of surface or subsurface rock, trenches up to 2m depth will be dug out all around the site. They will be filled with top soil from the construction site up to a depth of 1.75m leaving a gap on about 0.25m for effective watering and manuring. Where it is not possible for a trench; bore holes of 9 inches diameter and 10 Ft (3m) depth will be drilled, filled with a mixture of garden soil, vermicompost and sand in equal proportions. Saplings grown in poly bags will be transplanted. Fast growing, evergreen or semi evergreen tree saplings or cuttings will be planted to develop a thick green belt. Saplings are readily available from the local nurseries and the plants are well adapted to the local agro climate.

Shrubs and trees will be planted in encircling rows around the project site. The short trees (<10 m height) will be planted in the first rows (towards plant side) of the green belt. The tall trees (>10 m height) will be planted in the outer row (away from plant side). Some of guidelines to be considered are:

- Planting of trees in each row will be in staggered orientation.
- In the front row, shrubs will be grown.
- Since the trunks of the tall trees are generally devoid of foliage, it will be useful to have shrubs in front of the trees so as to give coverage to this portion.
- The spacing between the trees will be maintained slightly less than the normal spaces, so that the trees may grow vertically and slightly increase the effective height of the green belt.
- Providing the Greenbelt more than 33% area of the total project area with various species.

9.13 RAIN WATER HARVESTING

Current technologies of rainwater harvesting require that all roof and terrace water passes down into a covered tank where it can be stored for use after the monsoon. This is most advantageous in constructing large storage tanks which are expensive. Another way of using rooftop rainwater harvesting is to collect it so that it percolates into the ground to recharge wells instead of flowing over the ground into rivers. Thus, by recharging the groundwater by water harvesting from rooftops, the table rises and the surrounding wells retain water throughout the year.

For centuries, people have relied on rainwater harvesting to supply water for household, landscape, livestock, and agricultural uses. Before large centralized water supply systems were developed, rainwater was collected from roofs and stored on site in tanks known as cisterns. With the development of large, reliable water treatment and distribution systems and more affordable well drilling equipment, rain harvesting systems have been all but forgotten, even though they offer a source of pure, soft water. A renewed interest in this time-honoured technique/ approach emerged in recent decades due to following reasons:

- The problem of water shortage in arid and semi-arid regions
- The escalating environmental and economic costs of providing water by centralized water systems or by well drilling
- Health benefits of rainwater
- Potential cost savings associated with rainwater collection systems.

Rainwater harvesting is defined as a method for inducing, collecting, storing and conserving local surface runoff for later use. Three types of water harvesting are covered by rainwater harvesting:

1. Water collected from roof tops, courtyards and similar compacted or treated surfaces is used for domestic purpose or garden crops, or for groundwater recharge.
2. Micro-catchments water harvesting is a method of collecting surface runoff from a small catchments area and storing it in the root zone of an adjacent infiltration basin. The basin is planted with trees, bushes or with annual crops.
3. Macro-catchments water harvesting, also called harvesting from external catchments is the case where runoff from hill-slope catchments is conveyed to the cropping area located at hill foot on flat terrain.

Rain Water Harvesting is collecting and using precipitation from a catchments surface. The rainwater collected can be stored for direct use or used for hydrant purpose on emergency or can be recharged into the groundwater. Rain Water harvesting is thus becoming essential as it helps to meet our demand for the water either directly as rain water is stored and put to different uses or indirectly via replenishing the ground water. There are two main techniques of Rain Water Harvesting (RWH) (i) storage of rainwater on surface for further use (ii) Recharge to groundwater.

For RWH, the rainwater thus as it falls on a catchment (The surface that receives rainfall directly. It can be a paved area like a roof, terrace or courtyard of a building, or an unpaved area like a lawn or open ground) is carried through conduits (The pipelines or drains that carry exclusive rainwater from the catchment) for Storage or Recharge.

In general, runoff (It is the term applied to the water that flows away from a catchment after falling on its surface in the form of rain) from only paved surfaces (for example rooftop) is used for storing since it will be relatively less 'contaminated. Drainpipes that collect water from the rooftop are diverted to the storage container. Rainwater collected from rooftops is free of mineral pollutants like fluoride and calcium salts which may be found in groundwater. But it is likely to be contaminated with pollutants in air and on surface. To prevent leaves and debris form entering the system, mesh filters should be provided at the mouth of the drainpipe. It is preferable to use the stored water for purposes like gardening, flushing, bathing, augmenting storage for fire-fighting etc and even for utilities after necessary treatment If it is to be used for drinking purposes it should be used after boiling or disinfecting with chlorine tablets or after filtration.

9.14 REMOTE SENSING AND GIS

Geographical Information System (GIS) and Remote Sensing, play an important role in generating automated spatial datasets and in establishing spatial relationships. The Environmental Impact Assessment (EIA), for any project like dam construction on the Man river, Gujarat, India, was performed using GIS and Remote Sensing software-Arc/Info and ERDAS Imagine. Efficient management of irrigation water can be suggested using recent information of the command area by processing imageries through ERDAS Imagine. The impact of the dam in terms of catchment area and command area was computed to assess the net benefit to the society. GIS can also help in the site selection for the rehabilitation and infrastructure location.

Geospatial technology consists of remote sensing and GIS is an essential component of the Environmental Impact Assessment (EIA) process, as environmental resources are directly affected by changes in the shape and extent of the proposed disturbance. With the use of geo-spatial techniques like remote sensing, Geographical Information Systems (GIS), and Global Positioning Systems (GPS), EIA has enhanced substantial viewing, movement, query, and even map-making capabilities. However, one of the main challenges is to have access to the most up-to-date and accurate geospatial data and interpretations. With an emphasis on using geospatial data in particular, the value of the information resource is far higher than is generally available with text and numeric information. Several specific relevant applications of geospatial tools to integrate EIA are presented in the context of an Indian scenario. Applications have included monitoring of natural resources (air, water, land, etc.), ground-level ozone, soil erosion, study of sea-level rise due to global warming, change-detection studies, delineation of ecologically sensitive areas using digital-image analysis and Geographic Information Systems. This study focuses on the possibility of using a proposed spatial-decision-support system to conduct EIA, which should make it possible to upload, evaluate, maintain, and report field and analytical data that have been stored in a variety of formats.

The EIA is used: (1) to ensure that local agencies carefully consider significant environmental impacts arising from projects under agency jurisdiction; (2) to establish a procedure by which the public is given an opportunity for meaningful participation in the agency's consideration of the proposed action; and (3) to provide records for efficient spatial analysis. The EIA was designed to be a detailed and quantitative investigation which rigorously analysed the findings of potential environmental impact of the proposed project and also addressed the public concerns through the use of remote sensing and GIS technologies.

9.14.1 Geospatial Requirements

The inherent spatial requirements of an EIA, (that is the need to assess the impact of a proposed project on spatial analysis provides significant opportunities to apply GIS analysis for completing the EIA project. GIS analysis can greatly enhance the evaluation of EIA-required items. A case study of the use of GIS analysis for land, territorial resources, for example, land use, infrastructures and analysis of emissions and dispersion modelling system, like metereological and air pollution data is also analysed along with a discussion of the benefits of novel geomatic applications. Addressing land use and territorial resources require the quantification of the land to be affected by the project. Specifically, morphological and land use analysis requires estimating the acreage of various types of land use to be impacted. Spatial analysis requires estimating the amount of available data for EIA. Spatial multicriteria decision problems typically involve a set of geographically-defined alternatives or events from which a choice of one or more alternatives is made with respect to a given set of evaluation criteria (Jankowski, 1995; Malczewski, 1999).

9.14.2 Spatial Decision Support System (SDSS)

Two considerations are of paramount importance for spatial multicriteria decision analysis: (1) the GIS component such as data acquisition, storage, retrieval, manipulation, and analysis capability; and (2) the Spatial analysis component such as aggregation of spatial data and decision makers' preferences into discrete decision alternatives (Carver, 1991; Jankowski, 1995).

Densham (1991) lists the distinguishing capabilities and functions of Spatial Decision Support Systems (SDSS), which should be capable of: 1) providing mechanisms for the input of spatial data; 2) allowing representation of the spatial relations and structures; 3) including the analytical techniques of spatial and geographical analysis; and 4) providing output in a variety of spatial forms, including thematic cartography. SDSS typically have three components: a database management system and geographical database, a model-based management system (analytical modelling capabilities and analysis procedures), and a dialogue generation and management system (a user interface with display and report generators).

9.14.3 GIS in EIA

The development of Geographical Information Systems (GIS) analysis can serve as a valuable tool for EIA and Spatial Analysis. Geographical Information System (GIS) are computer systems that can store, integrate, analyze and display spatial data. The first systems evolved in the late sixties, and by mid seventies they have been used for EIA. In 1972 a computerised version of the technique was used for sitting power lines and roads (Munn, 1975). It is noteworthy that the so called "first GIS" (Canada GIS or CGIS) was used for EIA in the late 1970s for the preparation of an EIS for a dam on the river Thames (Griffith, 1980). GIS offers a special environment for dealing with the spatial properties of a project. Those special attributes of the GIS are very important for the analysis of environmental issues, since most of them are spatial by nature, and no other computerised system can handle them properly (Schaller,1990).

In recent years two important developments have helped in reduce the complexity of spatial analysis. In the last decade, due to the evolution of computer technology, and especially their graphic capabilities, GISs have become more user friendly and powerful. In addition the availability and quality of digital spatial data sets improved, to the level where they are now adequate for routine analysis. These two trends make possible the set up of and use of GIS at lower cost in terms of time and money than ever before. However, the use of GIS in EIA process in general, and for scoping in particular has been limited, due in part to their cost in terms of time and money relative to the time and budgets allocated for EIA preparation, and especially for scoping. Surveys of the use of GIS in EIA found that while GIS is widely utilised, its use is largely limited to the basic GIS functions such as map production, classic overlay or buffering. This utilisation does not make use of the key advantage of GIS for EIA, its ability to perform spatial analysis and modelling. Some of the uses of GIS for EIA are complex modelling representation techniques, repository of data and cumulative impact assessment. Spatial data analysis of data in GIS systems temporal variations and change detection analysis, creation of maps of with mandatory buffers. Such economies may be of particular relevance for the use of many GIS systems are insufficiently accurate for legal purposes due to several reasons, such as: limitations of the photogrammetric process; errors in the process of digitising existing maps; inaccuracies inherent in the maps; maps of different scales; different levels of cartographic representation and cartographic generalisation; and so on. Hence, the user of GIS for EIA study should be cautions in view of the above limitations. In an EIA framework, GIS can prove particularly useful for the evaluation of cumulative impacts. Smit and Spalding stress the potential of GIS for this type of analysis, arising from the ability to consider the spatial component and to allow the analysis of the temporal evolution (Smit et al., 1995).

REVIEW QUESTIONS

1. Listout the five technologies for sewage treatment
2. What are the advantages of biogas treatment plant?
3. Write a notes on design criteria of STP
4. Explain UASB and TSU-system
5. What do you mean by Root Zone Treatment
6. Write down the advantages of FAB system
7. Explain the advantages of FAB system
8. Explain (i) Removal of suspended solids from waste water (ii) Removal of dissolved solids
9. Explain the removal of dissolved organic compounds
10. What do you understand the air pollution control technologies
11. Explain Soil pollution control technology
12. Explain Phytoremediation techniques
13. Explain Biotemediation techniques
14. Explain Hazardous waste treatment technologies
15. Write a story on greenbelt development.
16. What do you understand rainwater harvesting?
17. Explain the role of RS and GIS in EIA.

CHAPTER 10

TOWARDS SUSTAINABLE FUTURE

10.1 CONCEPT OF SUSTAINABLE DEVELOPMENT

A system or process is sustainable if it can be continued indefinitely, without depleting any of the material and energy resources required to keep it running. The term was first applied to the idea of sustainable yields in human endeavors such as forestry and fisheries. Trees, fish, and other biological species normally grow and reproduce at rates faster than that required just to keep their populations stable. This built-in capacity allows every species to increase or replace a population following some natural disaster. Thus, it is possible to harvest a certain percentage of trees or fish every year without depleting the forest or reducing the fish population below a certain base number. As long as the number harvested stays within the capacity of the population to grow and replace itself, the practice can be continued indefinitely. The harvest then represents a sustainable yield. It becomes unsustainable only when trees are cut or fish are caught at a rate that exceeds the capacity of their present population to reproduce and grow. The concept of a sustainable yield can also be applied to freshwater supplies, soils, and the ability of natural systems to absorb pollutants without being damaged. The surrounding environment.

The notion of sustainability can be extended to include ecosystems. Sustainable ecosystems are entire natural systems that persist and thrive over time by recycling nutrients and maintaining a diversity of species in balance and by using the Sun as a source of sustainable energy. Sustainability. Stewardship and science are three unifying strategie themes and concepts or ideals that can move societies towards a sustainable future (Fig. 10.1).

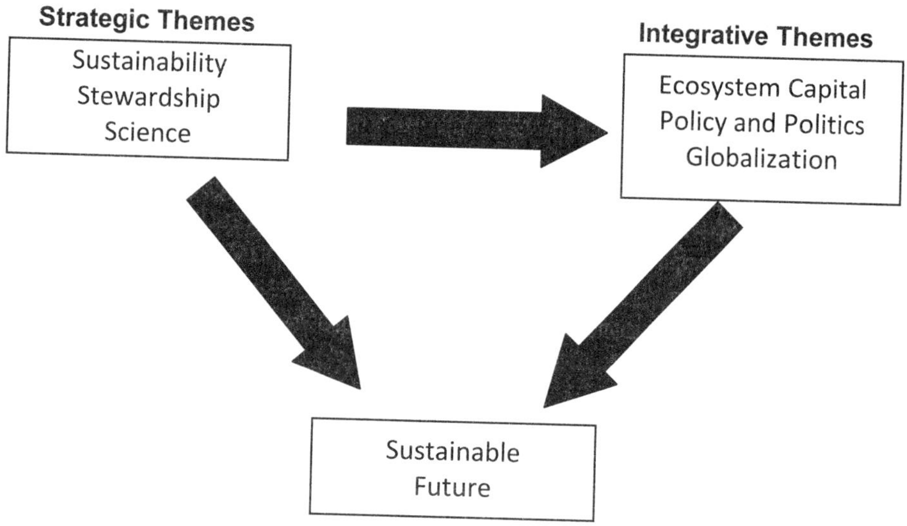

Goal

Fig. 10.1 Unifying themes for sustainability

Many of our interactions with the environment are not sustainable, however, demonstrated by such global trends as the decline of essential ecosystems and the increased emissions of greenhouse gases. Although population growth in the industrialized countries has almost halted, these countries are using energy and other resources at unsustainable rates, producing pollutants that are accumulating in the atmosphere, water, and land. In contrast, developing countries are experiencing continued population growth, yet are often unable to meet the needs of many of their people in spite of heavy exploitation of their natural resources. On the basis of expectations of continued economic growth and progress, the crux of the problem is seen to be modern society's inexperience with sustainability.

There are many dimensions to sustainable development like environmental, social, economic, political and no societies today have achieved anything resembling it (Fig 10.2). Nevertheless, as with justice, equality, and freedom, it is important to uphold sustainable development as an ideal-a goal toward which all human societies need to be moving, even if we have not achieved it completely anywhere. For example, policies and actions in a society that reduce infant mortality, increase the availability of family planning, improve the air quality, provide more abundant and pure water, preserve and protect natural ecosystems, reduce soil erosion, reduce the release of toxic chemicals to. The environment, restore healthy coastal fisheries, and so on, are all moving that society in the right direction-toward a sustainable future. At the same time, these actions are all quite measurable, so progress in achieving sustainable development can be assessed. Communities and organizations are developing sustainable development indicators and goals to track their progress, such as the Environmental Sustainability Index which evaluates the ability of nations to protect the environment.

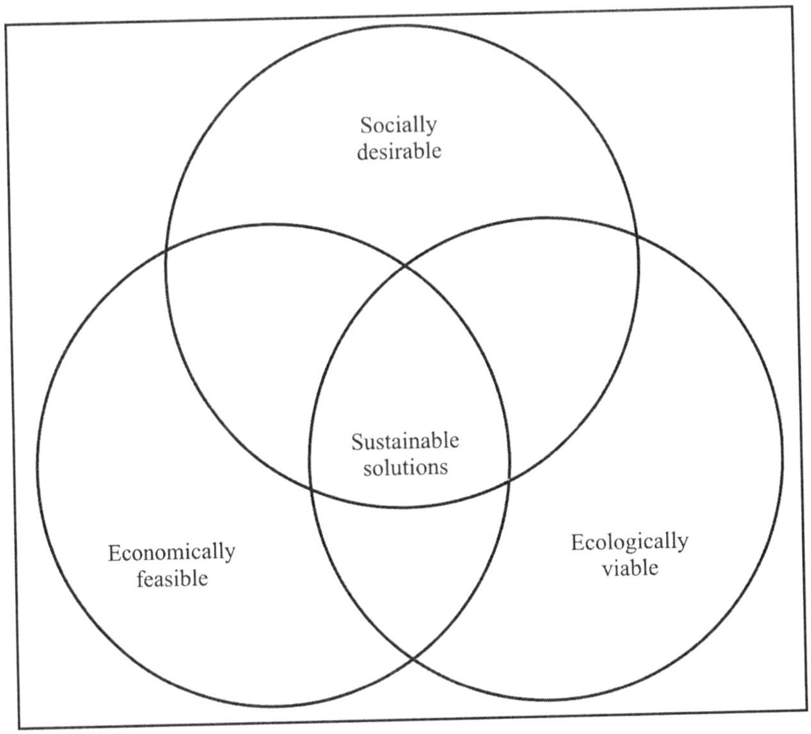

Fig. 10.2 Sustainable solutions. The concerns of sociologists, economists, and ecologists must intersect in order to achieve sustainable solutions in a society

10.2 THREATS TO SUSTAINABILITY

- Applications of sustainable solutions of developed countries to developing countries.
- To examine the dynamics of sustainability namely hierarchy of attributes, components and other systems at increasing scales at each significant level.
- Collapse of larger systems as crop declines.

Fertility and land use stresses have been only a tertiary factor. Sources of resilience that have buffered households and larger systems from the impacts of these range from farmers' strategies of crop and income diversification and searches for resistant cultivars to interventions by national and international bodies. Numerous population and large-scale system collapses have also occurred almost all caused by extreme social and/or biophysical perturbations. Increased land use pressure has led to the disappearance or alteration of many prior practices, though farmers have generally adapted to these.

10.3 HUMAN POPULATION AND ENVIRONMENT

The population is currently growing at the rate of 76 million people per year. This rate is equivalent to adding to the world every year according to the latest servey reports the population growth is occurred as medium projection predicts that population will pass the 7 billion mark in 2012, trillion mark in 2024, and the 9 billion mark in 21 will reach 9.1 billion in 2050. (Fig. 10.3).

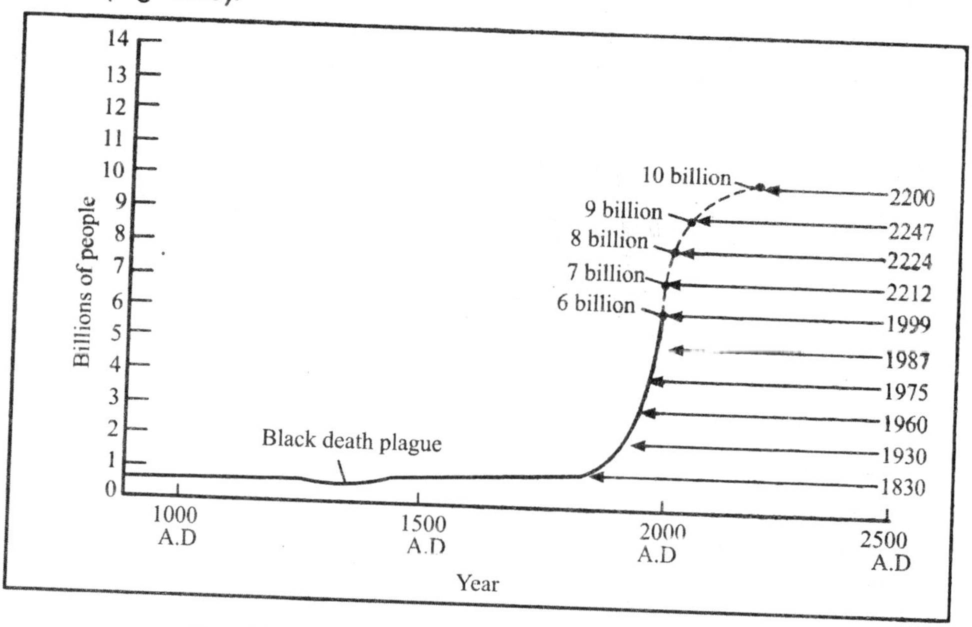

Fig. 10.3 World population over the centuries.

A population is the number of individuals of a given species within a specified area or habitat. In ecology, the study of population dynamics forms an important sub-field known as population biology. Population studies have been used in projecting the population has been an issue of debate. Scientists predict that human populations grow at faster rate lead to catastrophe due to which billions of people die. Center scholars think that current trends may no lead to disaster and find a reasonable and stable population level.

Populations increase through reproduction but they do not increase infinity because of environmental limitations including disease, predation, confection for space and nutrients and nutrient shortage. These various limit to population growth are collectively concept of this resistance is the carrying capacity i.e., the maximum number of individuals a habitat can support. When population exceeds its environment's carrying capacity, nutrients shortages may weather the individuals, reduce reproduction, and increase death rates, until the population falls below its maximum size of population. Its environment's carrying capacity, it is said, to overshoot its environment's capacity. A catastrophe usually follows an overshoot to disease the population well below. Its maximum. In many cases populations undergo repeated overshoot-dieback cycles, Sometimes these cycles gradually decrease until a stable population is reached with equilibrium in carrying capacity. As the population grows, the number of breeding adults increases so that growth accelerates. Increases at a constant accelerating rate of change is known as exponential growth. Generally more stable populations are desirable.

10.3.1 Population Explosion

The rapid growth of population is perhaps the most obvious factor affecting the present and future national and regional development, but it is by no means the only population problem in the world today. The main problem is that of 'Population Explosion'. Population explosion is a sudden spurt in the rate of population growth that occurs in the second stage of demographic transition due to a rapid fall in the mortality rate without a corresponding fall in the birth rate.

According to Theory of Demographic Transition, every country paşses through three stages of demographic transition. In the First stage, both birth and death rates are high and hence population remains more or less stable. The main features of this stage are backward economy where agriculture is the main occupation, low per capita income, low standard of living, inadequate and unbalanced diet, absence of educational opportunities and presence of social evils like polygamy, early marriages etc. In the second stage, there is rapid growth of population. With the beginning of process of development, the living standards of people improve, education expands/ medical and health facilities increase as a result of which death rate comes down. But as long as society remains primarily agrarian and education remains confined to a narrow section of the society, attitude of the people towards the size of the family not change radically and the birth rate remains high. In this situation, population increases at an alarming rate. If in a country the process of industrialization accompanied by urbanization is not fast and education is not widespread, then this is really a grave situation. This is called Population Explosion.

The effects of population explosion are numerous with far reaching consequences. Some of them are unemployment, low living standard of people, hindrance in the process of development of economy, pressure on agriculture land, low per capita income, lack of basic amenities like water supply and sanitation, education, health, etc., high crime rate, environmental damage, migration to urban area in search of job, energy crisis, Overcrowding of cities leads to development of slums.

10.4 OVER-EXPLOITATION OF RESOURCES

Overexploitation of natural resources through unsustainable hunting, fishing, or extracting raw material has serious implications for biodiversity. Besides reducing the number of species, it also diminishes diversity within the genetic pool. Decreasing species abundance impacts more than just the individual species: it disrupts the interaction between species in the ecosystem.

If renewable resources are harvested at a rate greater than their regeneration rate, the long-term flow of benefits is reduced, and they are said to be over harvested. When natural capital is drawn down too far, fundamental ecosystem changes can occur which make ecosystem recovery to full service delivery potential very slow or impossible, and degradation is said to have occurred. Degraded ecosystems support half or less of the biodiversity of non-degraded used ecosystems. Over harvesting is a problem in many localities. For example, about 9 percent of rangelands south of the equator are grazed by domestic livestock at unsustainable rates. The fish stocks in the Great Lakes show classic symptoms of over fishing, Over-exploitation in some cases lead to exhaustion, particularly by excessive forestry, fishing and hunting. This over-exploitation may be explained in part by human overpopulation in some areas of the planet, ever-increasing world demand for these resources and the development of international trade. Sustainable use of natural resources is essential to ensuring that future generations have the same quality of life that we have today.

10.5 STRATEGIES FOR ACHIEVING SUSTAINABLE DEVELOPMENT

The Rio Summit established sustainable development as the guiding vision for the development efforts of all countries. The strategies for sustainable development to ensure socially responsible economic development while protecting the resource base and the environment for the benefit of future generations. To agree how the international community can best assist developing countries in meeting good of sustainability. Commitment to provide support for sound nationally-owned sustainable development strategies where conditions for effective partnership are in place. In simple terms, sustainable development means integrating the economic, social and environmental objectives of society, in order to maximize human well-being in the present without compromising the ability of future generations to meet their needs. This requires seeking mutually supportive approaches whenever possible, and making trade-offs where necessary. For developing countries, and for development co-operation, reducing poverty and meeting the International Development Goals are imperatives — within the broad context of sustainable development

There has been progress in developing countries in some key areas, including reducing levels of absolute poverty and increasing attention to environmental issues. In other areas there has been significant deterioration, including water, soil, and biodiversity. Extreme poverty still ravages the lives of a quarter of the population in developing countries. Such levels of poverty and inequality raise serious issues of sustainability of peace and security, of equity and solidarity, and of the environment — at the national, regional and global levels.

10.5.1 Strategy Formulation

- Country ownership and participation, leadership and initiative in developing their strategies.
- Broad consultation, including particularly with the poor and with civil society, to open up debate on new ideas and information, expose issues to be addressed, and build consensus and political support on action.
- Ensuring sustained beneficial impacts on disadvantaged and marginalized groups and on future generations.
- Building on existing strategies and processes, rather than adding additional ones, to enable convergence and coherence.
- A solid analytical basis, taking account also of relevant regional issues, including a comprehensive review of the present situation and forecasts of trends and risks.
- Integration of economic, social and environmental objectives

- through mutually supportive policies and practices and the management of tradeoffs.
- Realistic targets with clear budgetary priorities.

The environment is a key determinant of growth and of poverty reduction. Environmental issues, including longer-term and global perspectives, need to be integrated into mainstream planning processes affecting these and other development objectives. We welcome the international discussions on the synergies between, and potential for convergence in, the underlying principles of country level planning frameworks – such as the poverty reduction strategy papers (PRSPs); the Comprehensive Development Framework (CDF); the National Visions and the National Action Plans. There is a particular opportunity to promote the better integration of environmental and other issues of sustainability into poverty reduction strategies, and we endorse the move to develop these as long-term sustainable poverty reduction strategies. More broadly, convergence is necessary to avoid duplication, confusion and straining of developing country capacity and resources. We endorse the emerging consensus on convergence and we recognize that putting this consensus into practice effectively is a learning process.

10.6 ENVIRONMENTAL EDUCATION

Environmental education (EE) refers to organized efforts to teach about how natural environments function and, particularly, how human beings can manage their behavior and ecosystems in order to live sustainably. The term is often used to imply education within the school system, from primary to post-

secondary. However, it is sometimes used more broadly to include all efforts to educate the public and other audiences, including print materials, websites, media campaigns, etc. Related disciplines include outdoor education and experiential education.

Environmental education is a learning process that increases people's knowledge and awareness about the environment and associated challenges, develops the necessary skills and expertise to address the challenges, and fosters attitudes, motivations, and commitments to make informed decisions and take responsible action (UNESCO, Tbilisi Declaration, 1978).

Environmental education generally refers to curriculum and programs which aim to teach people about the natural world and particularly about ways in which ecosystems work. Environmental education programs often aim to change people's perceptions about the value of the natural world and to teach how to change environmental behavior, such as getting people to recycle or how to build eco-friendly dwellings. Environmental education programs often aim to (i) help students develop factual knowledge about the natural environment, particularly with regard to how ecosystems work and human impacts on the natural environment; (ii) foster more positive perceptions about the value of the natural world (iii) develop eco-friendly habits, (iv) engage students in environmental rejuvenation projects and (v) develop students' psychological and spiritual relationship with nature. The components of environmental education are:

(i) Awareness and sentivity about the environment and environmental challenges
(ii) Knowledge and understanding about the environment and environmental challenges
(iii) Attitude concern for the environment and help to maintain environmental quality
(iv) Skills to mitigate the environmental problems
(v) Participation for exercising existing knowledge and environmental related programmes

Environmental Education helps ensure the health and welfare of any Nation by the following basic environmental concerns. The environmental concerns are : protecting human health, advancing quality education, creating jobs in the environmental field, promoting environmental protection along with economic development, encouraging stewardship of natural resources

10.6.1 Protecting Human Health

The link between environmental challenges and human health is a major cause of public concern about the environment. Lead poisoning from paint and pipes, air pollution, pesticides in water and food supplies, increased threats of skin cancer from depletion of the ozone layer, and other environmental and health issues are of growing concern to Americans, especially effects on children and future generations. Environmental education can help to prevent or mitigate environmental human health problems by providing the public with information on the causes of environmental pollution. It also gives knowledge on how pollutants may affect health, how to assess real versus exaggerated risks, and how to make informed and responsible decisions that prevent or mitigate the effects of pollution on health.

10.6.2 Advancing Quality Education

Educators and public officials generally believe that improvements are needed in the nation's public education system to enhance student learning. Many educational scholars and practitioners agree that students are not doing well at thinking, reasoning, analyzing, or problem solving. Many goals of education reform emphasize the need to strengthen core subjects such as maths, science, and geography; teach across subject areas; improve critical- thinking and problem-solving skills; and relate learning in the classroom to the needs and issues of the community. Environmental education has a tremendous potential to contribute to these goals of education reform. For example, it can strengthen teaching in many core subjects, especially science, because science is the basis for solving many of our environmental challenges. At the same time, environmental education can promote interdisciplinary teaching because environmental topics can be addressed from many different perspectives, including scientific, historical, and cultural. Environmental education can also bring local environmental challenges into the classroom to improve analysis and problem- solving skills.

The final aspect of environmental education policies, but certainly not least important, is training individuals to thrive in a sustainable society. In addition to building a strong relationship with nature, American citizens must have the skills and knowledge to succeed in a 21st century workforce. Thus, environmental education policies fund both teacher training and worker training initiatives. Teachers must be trained to effectively teach and incorporate environmental studies in their curricula. On the other hand, the current workforce must be trained or re-trained so that they can adapt to the new green economy. Environmental education policies that fund training programs are critical in educating citizens to prosper in a sustainable society.

10.7 CONSERVATION OF RESOURCES

Management of the human use of natural resources to provide the maximum benefit to current generations while maintaining capacity to meet the needs of future generations. Conservation includes both the protection and rational use of natural resources. Earth's natural resources are either nonrenewable, such as minerals, oil, gas, and coal, or renewable, such as water, timber, fisheries, and agricultural crops. The combination of growing populations and increasing levels of resource consumption is degrading and depleting the natural resource base. Nonrenewable resources, such as fossil fuels, are replaced over geologic time scales of tens of millions of years. Human societies will eventually use up all of the economically available stock of many nonrenewable resources, such as oil. Conservation entails actions to use these resources most efficiently and thereby extend their life as long as possible. By recycling aluminum, for example, the same piece of material is reused in a series of products, reducing the amount of aluminum ore that must be mined. Similarly, energy-efficient products help to conserve fossil fuels since the same energy services, such as lighting or transportation, can be attained with smaller amounts of fuel.

It may be expected that the biggest challenge of resource conservation would involve nonrenewable resources, since renewable resources can replenish themselves after harvesting. In fact, the opposite is the case. Historically, when nonrenewable resources have been depleted, new technologies have been developed that effectively substitute for the depleted resources. Indeed, new technologies have often reduced pressure on these resources even before they are fully depleted. Fiber optics, for example, has substituted for copper in many electrical applications, and it is anticipated that renewable sources of energy, such as photovoltaic cells, wind power, and hydropower, will ultimately take the place of fossil fuels when stocks are depleted. Renewable resources, in contrast, can be seriously depleted if they are subjected to excessive harvest or otherwise degraded, and no substitutes are available for, say, clean water or food products such as fish or agricultural crops. Moreover, when the misuse of biological resources causes the complete extinction of a species or the loss of a particular habitat, there can be no substitute for that diversity of life. Critical elements of the effective conservation of natural resources include sustainable resource management, establishment of protected areas, and ex situ (off-site) conservation.

10.7.1 Resource Management

Some of the most pressing resource conservation problems stem directly from the mismanagement of important biological resources. A number of steps are being taken to improve resource conservation in managed **ecosystems.** (1) Considerable scientific research has been undertaken to better understand the natural variability and productivity of economically important resources. (2) Many national and local governments have enacted regulations for resource management practices on public and private lands. (3) In some of regions, programs recently have been established either to involve local communities in resource management decisions or to return to them resource ownership rights. (4) Efforts to manage resources on a regional or ecosystem scale using methods that have come to be known as ecosystem management or bioregional management. Since the actions taken in one location often influence species and processes in other locations, traditional resource conservation strategies were often focused too narrowly to succeed.

10.7.2 Protected areas

One of the most effective strategies to protect species from extinction is the establishment of protected areas designed to maintain populations of a significant fraction of the native species in a region. Worldwide, 9832 protected areas, totaling more than 9.25 million square kilometers (24 million square miles), cover about 8% of land on Earth. Although these sites are not all managed exclusively for the conservation of species, they play an essential role in protecting species from extinction. Several regions with important biodiversity still lack effective protected-area networks. In addition, where protected areas have been designated, human and financially resources are not always available to effectively manage the areas. Particularly in developing countries like India. The establishment of protected areas has resulted in conflicts with local communities that had been dependent upon the areas for their livelihood.

10.7.3 Ex situ conservation

The most effective and efficient means for conserving biological resources is to prevent the loss of important habitats and to manage resources for their long-term productivity of goods and services. In many cases, effective conservation in the field is no longer possible. For example, some species have been so depleted that only a few individuals remain in their natural habitat. In these cases, there is no alternative to the ex situ conservation of species and genetic resources in zoos, botanical gardens, and seed banks. Ex situ collections play important conservation roles as well as serving in public education and research.

10.8 URBAN SPRAWL

Sprawl's signature manifestation is a far-flung urban-suburban network of low-density residential areas, shopping malls, industrial parks, and other facilities loosely laced together by multilane highways. The word sprawl is used because the perimeters of the city have simply been extended outward into the countryside, one development after the next, with little plan as to where the expansion is going and no notion as to where it will stop. Almost everywhere we go near urban areas, we are confronted by farms and natural areas giving way to new developments, new highways being constructed, and old roadways being upgraded and expanded. Cities had developed in ways that allowed people to meet their needs by means of the transportation that was available-mainly walking. Every few blocks had a small grocery, a pharmacy, and other stores, as well as professional offices integrated with residences. Often, buildings had a store at the street level and residences above. Schools were scattered throughout the city, as were parks for outdoor recreation or relaxation. Thus, walking distances were generally short, and bicycling made going somewhere even more convenient. For more specialized needs, people boarded public transportation-electric trolleys.

Despite the many advantages of urban life, many people found cities less-than-pleasant places to live. Especially in industrial cities, poor housing, inadequate sewage systems, inadequate refuse collection, pollution from home furnaces and industry, and generally congested, noisy conditions were common. A decrease in services during World War II aggravated those problems.

10.8.1 Urbanization

Urbanization is a part of the development process urbanization process is fast in rapidly growing economics where newly established industries and ancillary activities continuously provide jobs to people who with to migrate to cities. The economic full of cities becomes strong if industrial growth is fast and in spite of high capital intensity industries offer jobs in increasingly large number. The process of urbanization slows down only when the proportion of urban population to total population in a country becomes very high.

For measuring urbanization a number of measures are being used. One very simple method is to examine the changes in the level of urban population. If we employ this method we observe that in 1991, the urban population of India was a little more than ¼ of total population , as against 1/6 in 1951. this clearly indicates that the urbanization process has been stable. The urban rural growth differential (URGS) method also indicates the same tend,. Since urban and rural population growth rates in India are presently very close to each other, this measure provides a good idea of the rural — urban transformation. A third method is to examine the growth of urban population it self using this criteria we notice that while the total population in two country since 1951 has since to about 3.6 times, the urban population in the same period close to 10.0 times.

Towns are classified into the following six classes according to population size.

Class-I	Population of 100,000 and above
Class-II	Population of 50,000 - 99,999
Class-III	Population of 20,000 - 49,999
Class-IV	Population of 10,000 - 19,999
Class-V	Population of 5,000 - 9,999
Class-VI	Population of less than 5,000

Population in India had tended to concentrate in Class I towns (according to Indian census convention, all such towns are know as cities) Again, life in developed countries, in India, the maximum concentration of urban population is found in the largest urban centers within cities especially in cities with population more than ten lakes etc.. (Also known as million plus cities).

10.9 HUMAN HEALTH

Environment-related issues that affect our health have been one of the most important riggers in the increasing awareness of the need for better environmental management. The changes in our environment induced by human activities in nearly every sphere of life have had an influence on our health patterns. The assumption that the only indicator of human progress is economic growth is not true. We expect urbanization and industrialization to bring in prosperity, but on the downside, it leads to diseases related to overcrowding and poor quality drinking water, resulting in an increase in water-borne diseases like infective diarrhea and air-borne bacterial diseases like tuberculosis. High-density city traffic leads to an increase in respiratory diseases like asthma. Agricultural pesticides that enhanced food supplies during the green revolution have affected both the farm worker and all of us who consume the produce. Modern medicine promised to solve many health problems, especially associated with infectious diseases through antibiotics, but bacteria have found ways to develop resistant strains, frequently even changing their behavior in the process, making it necessary to keep on creating newer antibiotics. Many drugs have been found to have serious side-effects. Sometimes, the cure is as damaging as the disease process itself.

Thus, development has created several long-term health problems. While better health care has led to longer life-spans, coupled with lowered infant mortality, it has also led to an unprecedented growth in our population which has negative implications on environmental quality. A better health status of society will bring about a better way of life only if it is coupled with stabilizing population growth.

10.9.1 Human Rights

The term 'Human Rights' refers to those basic rights which are essential for the development of human personality such as right to life, liberty, property and security of an individual. The 'Universal Declaration of Human Rights' adopted by the United Nations on December 10,1948, has served as a Magna Carta for all humanlity. It represented a 'world milestone in the long struggle for human rights'.

The declaration states that—"the inherent dignity of all members of the human family is the foundation of freedom, justice and peace in the world". This is possible only when each and every human being enjoys fundamental rights, which include:

- the right to life, liberty and security of person;
- the right to own property;
- the right to freedom of opinion and expression;
- the right to an adequate standard of living;
- the right to seek and to enjoy in other countries asylum from persecution;
- the right to education, freedom of thought, conscience and religion; and
- The right to freedom from torture and degrading treatment, etc.

This problem is very much under control by making the female education free at the primary level by the several underdeveloped nations. In India too the female literacy rate has gone upto 50% from 39.3%. Since independence, India has tripled its literacy and female literacy has increased by five times. Also certain income generating activities are also created to improve the socio-economic status of women in rural areas as well as in urban areas.

For the welfare and development of children, the Integrated Child Development Services (ICDS), aims to provide an integral package of services of regular health checkups, immunization, supplementary nutrition, referral services, pre-school education; and nutrition and health education to children, mothers, and pregnant women. Apart from ICDS, the other programmes for child development and welfare are— day-care centres for children below five years belonging to weaker sections, Balwadi Nutrition

Programme, Early Childhood Education and National Institute of Public Cooperation and Child Development. The Children Act, 1960 in India (amended in 1977) provides for the care, maintenance, welfare, training, education, and rehabilitation of the delinquent child. It covers the neglected and destitute, socially handicapped, uncontrollable, victimized and delinquent children. Article 39 (f) of the Constitution of India provides that "the state shall in particular direct its policy towards securing that childhood and youth are protected against moral and material abandonment".

10.9.2 Value Education

Values are not only with beliefs but also with our understanding, feelings and behavior. Value education may be understood in a broad sense to mean all aspects of the process by which teachers transmit values to the students.

Ancient Indian education had always given importance to value components. The education up to eighteenth century involved study of scriptures and Upanishads which are essentially the discourses on morality, ethics and virtuosity. The value education was the pivotal concern around which other knowledge areas were developed. Later the influences of cultures from across the boundaries and British colonization brought about changes in the structure of society. Urban living replaced rural symbiotic existence; joint family gave way to micro family and consequent changes sneaked into the educational system too. Apart form these developments, simultaneously growing influence of science and technology and consequent industrialization led to efforts aimed at strengthening of the education system by increasing inputs on science and technology. The efforts were successful. Green revolution, white revolution, self-sufficiency in steel production, etc. and now spearheading the information technology are the supporting evidence. However, in this process of modernization of education the emphasis shifted to economic and secular values. The education system lost its role as the guardian of ethical and moral values. The efforts are continuing to reconstruct education to fulfill its role in preservation of the moral character of the society and to keep pace with the economic and technological developments.

10.9.3 HIV / AIDS

The Human Immuno-deficiency Virus (HIV) causes Acquired Immunodeficiency Syndrome (AIDS) through contact with the tissue fluids of infected individuals, especially through sexual contact. As it reduces an individual's resistance to disease, it causes infected individuals to suffer from a large number of environment-related diseases and reduces the ability of infected individuals to go about their normal lives. It saps their strength, leads to skin lesions (karposi's sarcoma) and they become increasingly vulnerable to any air –or water-borne pollutant, until they eventually die.

In sub-Saharan Africa, where the infection has become highly prevalent, it is leading to great suffering and worsening poverty. These patients become unable to work and their usual source of incomes generation is lost. An increasing proportion of the poor are affected. Incomes lost due to the stigma of HIV / AIDS are usually met by the suffers by overexploiting their resource base, as they are not in any position to think of the long-term future. In Africa, this has led to degradation of the ecosystem and an increase of pressures from other impacts such as overuse of medicinal plants and poaching for wildlife. In South Africa, for example, people have a mistaken belief that turtle eggs can cure HIV /AIDS, thus leading to the eggs being over – harvested. When males die of the disease, agricultural work has to be taken over by already over-burdened women and children, affecting land management and productivity. Providing balanced diets and nutritional support for these poverty stricken patients can be partially addressed by better natural resource management like afforestation, access to clean water and wholesome food.

HIV / AIDS has a serious impact on the socioeconomic fabric of society. By 2002, India had an estimated 3.97 million infected individuals. There is a great need to organize AIDS education on prevention and management of the disease. This needs to be done through the formal educational sector and by using non-formal methods. Education is also important to reduce the stigma and discrimination against these patients. In India, women who are not socially empowered, are at a great disadvantage as they are powerless to demand safe sex from their partners. Women also have an added burden of caring for HIV – infected husbands. This produces enormous economic stresses on their family. HIV in India is rapidly moving from being a primarily urban disease to rural communities.

Research in Nepal has shown a linkage between rural poverty, deforestation and a shift of population to urban areas resulting in a rising number of AIDS patients. Prior to 1992, it was mainly seen in males who migrated to urban centers. In more recent times, a growing number of women are moving to Indian cities as sex workers. Women engaged in prostitution find it difficult to make partners take protective measures, such as the use of condoms that provide safe sex. A large proportion became victims of the diseases.

Blood transfusion from an infected person can also lead to HIV / AIDS in the recipient, as well as drug abuse by sharing needles with an infected person. However, the most important measure to prevent AIDS is the proper use of condoms that form a barrier to the spread of the virus during intercourse.

10.9.4 Role of Information Technology in Environment

Information generally has an advantage over those to whom such access is denied. The evolution of information technology, has generally led to dispersal of information to increasing broader audiences. Advances in telecommunications and other forms of information technology have contributed to the creation of new patterns of work and human association.

Information technology has also increased the pace of discovery. The capacity of establishing and maintaining worldwide databases has linked environmental, researches around the globe. The advancement in computer, communication, satellite and other technological developments have enabled engineers or environmentalists to gather relevant information simultaneously from many sources. These information is utilized for developing and early warning system and to forecast any eventuality much earlier. A large amount of information is easily available through Remote Sensing technology, Geographical Information System (GIS) and Global Positioning System (GPS) that is being used for various environmental studies.

Ministry of Environment, Government of India has established a Environmental Information System (ENVIS) in 1982. This has been established as a decentralized information system network whose purpose is collection, storage, retrieval and dissemination of environmental information to decision makers, policy makers, planners, scientists, engineers, environmentalists, researchers and general public all over the country. The ENVIS network has its focal point in the Environmental ministry along with thirteen subject oriented centers, known as ENVIS Centers, set up in the various institutions organizations of the country in the priority areas of environment like pollution control, toxic chemicals, energy and environment, environmentally sound and mangroves, corals and lagoons, media and environment etc.

New communications links are particularly vital to make use of such information sources as Geographic Information Systems (GIS), a computer-based system for gathering, manipulating, and analyzing environmental data. GIS databases are commonly established with information obtained from remote surveys via satellite and a variety of atmospheric and ground level surveys. GIS software packages and databases, which have almost unlimited applications, have a crucial relevance for national and local environmental management and learning.

GIS allows for the simulation of hypothetical models of environmental management and can demonstrate how subtle changes in one element of a landscape may have a powerful effect elsewhere.

Online healthcare information about Medicare and various related web sites guide consumers to a wide variety of health information available including the full text of clinical practice guidelines, and consumer brochures developed. A number of centers for disease control & prevention, are functioning and the public health service maintain their websites.

10.9.5 Women and Child Welfare

The Department of Women and Child Development was set up in the year 1985 as a part of the Ministry of Human Resource Development to give the much needed impetus to the holistic development of women and children. As the national machinery for the advancement of women and children, the Department formulates plans, policies and programmes; enacts/amends legislation, guides and coordinates the efforts of both governmental and Non-Governmental Organizations working in the field of Women and Child Development. Besides, playing its nodal role, the Department implements certain innovtive programmes for women and children. These programmes cover welfare and suport services, training for employment and income generations, awareness generation and gender senitization. These programmes play a supplementary and complementary role tot he other genera developmental programmes in the sectors of health, education, rural development etc. All these efforts are directed to ensure that women are empowered both economically and socially and thus become equal partners in national development along with men.

For the holistic development of the child, the Department has been implementing the world's largest almost unique and outreach programme of Integrated Child Development Services (ICDS) providing a package of services comprising of the following :

- Supplementary nutrition
- Immunization,
- Health check-up and referral services,
- Pre-school non-formal education.

There is effective coordinational and monitoring of various sectoral programmes. Most of the programmes of the Department are run through NGOs. Efforts are made to have more effective involvement of NGOs.

10.10 ENVIRONMENTAL ETHICS

Environmental Ethics deal with issues related to the rights of individuals that are fundamental to life and well-being. These concern not only the needs of each person today, but also those who will come after us. It also deals with the rights of others living creatures that inhabit the Earth.

In the 1970s, philosophers began to formulate a new field called 'Environmental Ethics'. Environmental ethics asks about the moral relationship between humans and the world around as; in contrast to traditional ethics, which concerns with relationship among people only.

The need of Environmental Ethics has arised as a result of the following three factors:

1. *New Effects on Nature :* The modern technological civilization has been affecting nature greatly; therefore, we should analyze the ethical consequences of human actions.
2. *New Knowledge about Nature :* Until a few decades back, only a small section of people realized that human's activities could be changing (or altering) the global environment. Now, the modern science demonstrates how humans have changed and are changing the global environment in ways not previously understood. For cxample, it has now been proved that burning of fossil fuels and deforestation have increased the carbon dioxide concentration in the atmosphere, and that this may lead to irreversible changes in global climate; Thus, new knowledge and better understanding of nature is raising new ethical issues.
3. *Expanding Moral Concerns :* Perhaps the most important question in environmental ethics is whether moral extensionism encompasses non-humans. Does nature have rights? Do other species have rights as well? Are they (other species) moral agents or at least moral subjects? Do we have a moral obligation to leave the environment in good conditionor our descendents, or are we at liberty to use environmental resources to the point of depletion within our life time? These expanded concerns lead to a need of environmental ethics.

10.11 ENVIRONMENTAL ECONOMICS

Environmental economics is a subfield of economics concerned with environmental issues. Quoting from the National Bureau of Economic Research Environmental Economics program: Environmental Economics undertakes theoretical or empirical studies of the economic effects of national or local environmental policies around the world. Particular issues include the costs and benefits of alternative environmental policies to deal with air pollution, water quality, toxic substances, solid waste, and global warming.

Central to environmental economics is the concept of market failure. Market failure means that markets fail to allocate resources efficiently. As stated by Hanley, Shogren, and White (2007) in their textbook *Environmental Economics*: "A market failure occurs when the market does not allocate scarce resources to generate the greatest social welfare. A wedge exists between what a private person does given market prices and what society might want him or her to do to protect the environment. Such a wedge implies wastefulness or economic inefficiency; resources can be reallocated to make at least one person better off without making anyone else worse off." Common forms of market failure include externalities, non excludability and non rivalry.

Environmental economics is related to ecological economics but there are differences. Most environmental economists have been trained as economists. They apply the tools of economics to address environmental problems, many of which are related to so-called market failures—circumstances wherein the "invisible hand" of economics is unreliable. Most ecological economists have been trained as ecologists, but have expanded the scope of their work to consider the impacts of humans and their economic activity on ecological systems and services, and vice-versa. This field takes as its premise that economics is a strict subfield of **ecology**. Ecological economics is sometimes described as taking a more pluralistic approach to environmental problems and focuses more explicitly on long-term environmental sustainability and issues of scale.

Environmental economics is viewed as more pragmatic in a price system; ecological economics as more idealistic in its attempts not use money as a primary arbiter of decisions. These two groups of specialists sometimes have conflicting views which may be traced to the different philosophical underpinnings.

Environmental economics was once distinct from resource economics. Natural resource economics as a subfield began when the main concern of researchers was the optimal commercial exploitation of natural resource stocks. But resource managers and policy-makers eventually began to pay attention to the broader importance of natural resources It is now difficult to distinguish "environmental" and "natural resource" economics as separate fields as the two became associated with sustainability. Many of the more radical green economists split off to work on an alternate political economy.

Environmental economics was a major influence for the theories of natural capitalism and environmental finance, which could be said to be two sub-branches of environmental economics concerned with resource conservation in production, and the value of biodiversity to humans, respectively. The theory of natural capitalism (Hawken, Lovins, Lovins) goes further than traditional environmental economics by envisioning a world where natural services are considered on par with physical capital.

The more radical Green economists reject neoclassical economics in favour of a new political economy beyond capitalism or communism that gives a greater emphasis to the interaction of the human economy and the natural environment, acknowledging that "economy is three-fifths of ecology" - Mike Nickerson.

10.12 THE CONCEPT OF GREEN BUILDINGS

Green building is a whole-systems approach for designing and constructing buildings that conserve energy, water, and material resources and are more healthy, safe, and comfortable. Many think of solar panels when they think of "green" building. The reality is that environmentally sustainable building goes far beyond energy consumption. Building materials and use of landfills during construction can have detrimental effects on volunteers, home owners and the environment. Green building offers a response to the realization that the way we have been building everything from houses to skyscrapers is not sustainable. Many health problems today stem from, or are aggravated by poor indoor air quality and exposure to toxic substances contained in commonly used building products. Green building practices can eliminate these health damaging conditions.

Adopting even one or two green strategies can have significant benefits for the home owner as well as for the environment,

- Energy efficiency is one of the primary advantages of green building. Energy consumption can be dramatically slashed. Below are a few of the strategies that go into making a house exceptionally energy efficient.

- Orient the house to reduce solar gain in summer and capture the sun's light and warmth in winter.
- Carefully sized overhangs or awnings will protect windows from the summer sun while admitting the sun's warming rays in winter when it is at a lower angle known as a ground-source heat pump system, consumes no fossil fuels at all, and provides outstanding performance year-round with an extraordinarily low operating cost.
- Maximize natural light to reduce the need for electrical usage during the day
- Compact fluorescent lights (CFL's) are big energy savers. Incandescent bulbs are highly inefficient, converting just 10% of the energy they use into light — the other 90% produces only heat. GEL's are up to six times more efficient and last up to ten times longer. Choose CFL's with warm color temperatures (around 2,7000 to 3,000° Kelvin) which are indistinguishable from incandescent lights.
- Cut energy consumption further with clean, renewable energy from photovoltaic panels. During periods when the panels produce more power than the house is using, the electric meter will actually run backwards. In some locales, wind generated electricity is also an option
- All newly built homes to produce more energy than they consumed by 2020. Renovate all existing buildings to save energy. Ban incandescent light bulbs by 2010. Reduce green house-gas emissions by 20% by 2020.
- Increase renewable energy from 9% to 20-25% of total energy consumptions by 2020.
- Bring transport emissions back to 1990 levels. Reduce vehicle speed limits by 10 kilometers per hour. Taxes and incentives to favour clean cars. Shift half of haulage by road to rail and water within 15 years. Develop rail and public transport.
- Reduce air pollutants quantitatively.
- Create a national network of green corridors and nature reserves.
- Increase organic farming from 2% to 6% of total acreage production by 2010 and to 20% by 2020.
- Ecological groups to be stakeholders, like trade unions, in government negotiations.
- Create a body to review planting of genetically modified crops on a case-by-case basis.

10.12.1 Design Considerations

The following are considered in designing green buildings:

Design an energy-efficient building: Use high levels of insulation, high-performance windows, and tight construction. In southern climates, choose glazings with low solar heat gain.

Design buildings to use renewable energy: Passive solar heatiiig, day lighting, and natural cooling can be incorporated cost-effectively into most buildings. Also consider solar water heating and photovoltaicor design buildings for future solar installations.

Optimize material use: Minimize waste by designing for standard ceiling heights and building dimensions. Avoid waste from structural over-design (use optimum-value engineering/advanced framing). Simplify building geometry.

Design water-efficient, low-maintenance landscaping: Conventional lawns have a high impact because of water use, pesticide use, and pollution generated from mowing. Landscape with drought-resistant native plants and perennial groundcovers.

Mote it easy for occupants to recycle waste: Make provisions for storage and processing of recyclables—recycling bins near the kitchen, under sink compost receptacles, and the like. Look into the feasibility of gray water: Water from sinks, showers, or clothes washers (gray water) can be recycled for irrigation in some areas. If current codes prevent gray-water recycling, consider designing the plumbing for easy future adaptation.

Design for durability: To spread the environmental impacts of building over as long a period as possible, the structure must be durable. A building with a durable style ("timeless architecture") will be more likely to realize a long life.

Design for future reuse and adaptability: Make the structure daptable to other uses, and choose materials and components that can be reused or recycled.

Avoid potential health hazards—radon, mold, pesticides: Follow recommended practices to minimize radon entry into the building and provide for future mitigation if necessary. Provide detailing to avoid moisture problems, which could cause mold and mild growth.

Renovate older buildings: Conscientiously renovating existing buildings is the most sustainable construction.

Create community: Development patterns can either inhibit or contribute to the establishment of strong communities and neighborhoods. Creation of cohesive communities should be a high priority.

Encourage in-fill and mixed-use development: In-fill development that increases density is inherently better than building on undeveloped (Greenfield) sites. Mixed-use development, in which residential and commercial uses are intermingled, can reduce automobile use and help to create healthy communities.

10.12.2 Green Building Movement in India

The Green Building movement has gained tremendous momentum during the past 3-4 years, ever since the Green Business Centre embarked on achieving the prestigious the Leadership Energy and Environmental Design rating for their own centre at Hyderabad. The "Platinum Rating" for the Green Business Centre building has sensitized the stakeholders of the construction industry. Today, several corporate and Government are considering Green buildings in a major way. This has resulted in a spurt in the demand for Green materials& equipment. The Indian materials and equipment manufacturers are now faced with the challenge to seriously look at Green features to meet the growing demand for Green Buildings.

Eco or green design principles are universal it cannot be one for USA, one for India and one for Japan. Most of the green building rating systems touch on the same chord that is constipation of resources. But the LEED rating system has turned out to be the most versatile and robust. After considering various rating systems, the Indian Green Building Council (IGBC) decided to adopt the LEED rating system. The IGBC is working in India to indigenise the LEED rating system to include the local factors. 'LEED India' rating which considers local Indian codes and standards is in an advanced stage of development.

The green building experiences in India have been exciting and challenging as well. Since its introduction in 2001, the LEED rating has emerged as a very useful tool in designing a green building. The LEED rating provided opportunities to introduce new products and materials. The easy availability of most of the green materials and equipment in the country has made it easier for the designers to adopt local materials to a very large extent. Now there is an imminent need for service providers, who would be required in large numbers, not in hundreds but thousands, as the movement is heading to reach greater heights. The green building movement is here to stay for the benefit of individuals, society and the country at large.

Green Building, also known as green construction or sustainable building, is the practice of creating structures and using processes that are environmentally responsible and resource- efficient throughout a building's life-cycle: from siting to design, construction, operation, maintenance, renovation, and deconstruction. This practice expands and complements the classical building design concerns of economy, utility, durability, and comfort.

Although new technologies are constantly being developed to complement current practices in creating greener structures, the common objective is that green buildings are designed to reduce the overall impact of the built environment on human health and the natural environment by:

- Efficiently using energy, water, and other resources
- Protecting occupant health and improving employee productivity
- Reducing waste, pollution and environmental degradation

10.12.3 Clean Development Mechanism (CDM)

The Clean Development Mechanism (CDM) was established under Article 12 of the Kyoto Protocol adopted by the third Conference of the Parties to the Framework Convention on Climate Change on December 11, 1997. The dual goals of the CDM are to promote sustainable development in developing countries and to allow industrialized countries to earn emissions credits from their investments in emission-reducing projects in developing countries. To earn credits under the CDM, the project proponent must prove and have verified that the greenhouse gas emissions reductions are real, measurable and additional to what would have occurred in the absence of the project.

As in many developing countries, the transportation sector is the largest source of energy- related carbon dioxide ($C0_2$) emissions. The e transportation sector accounted for 36 per cent of total CO_2 emissions from energy consumption, and transportation CO_2 emissions could increase by a factor of four by 2020 without mitigation measures.

The CDM represents a crucial opportunity to develop a sustainable solutions to enhance capacity and expand technology transfer opportunities. For this to be successful, it will be critical to have in place the technical and institutional capacity to develop, review and approve CDM projects in a systematic manner; assess and remove barriers to project implementation; and market its project opportunities effectively. The proposed project currently underway seeks to address these needs.

REVIEW QUESTIONS

1. Describe the unifying themes for sustainability.
2. Discuss the main dimensions to sustainable development.
3. Write short notes on (i) threats to sustainability (ii) human population (iii) the effects of population explosion (iv) over exploitation of resources
4. Explain various strategies for achieving sustainable development.
5. How do you narrate the advancing quality education?
6. Write an essay on Urban sprawl.
7. Discuss environmental ethics.
8. Discuss environmental economics.
9. Discuss the concept of Green Building and its design considerations.
10. Discuss the Green Building movement and LEED rating system.
11. Explain Clean Development Mechanism (CDM).
12. Write short notes on (i) Human health (ii) IT in Environmental problem solving.